D1577380

ANIMAL MIGRATION

First published in the UK by the Natural History Museum, Cromwell Road, London SW7 5BD

A catalogue record for this book is available from the British Library.

ISBN 978-0-565-09243-6

10 9 8 7 6 5 4 3 2 1

Originated in Hong Kong by Modern Age.
Printed and bound in Singapore by Star Standard Industries (Pte) Ltd.

Conceived, edited, and designed
in the United Kingdom by
Marshall Editions
The Old Brewery
6 Blundell Street
London N7 9BH
www.marshalleditions.com

Publisher Jenni Johns
New Titles Development Manager Deborah Hercun
Managing Editor Paul Docherty
Art Director Ivo Marloh
Picture Manager Veneta Bullen
Layout 3RD-I
Maps Mark Franklin
Indexer Sue Butterworth
Production Nikki Ingram

Below Wildebeest file across the sunburnt savanna of the Masai Mara in search of fresh grazing, with tails twitching constantly to keep biting flies at bay. **Previous page** Surfbirds gather on the mudflats of Alaska's Copper River delta. These shorebirds fly as far south as the Straits of Magellan for the winter.

ANIMAL MIGRATION

Remarkable journeys by air, land and sea

Ben Hoare

Published by the Natural History Museum, London

Contents

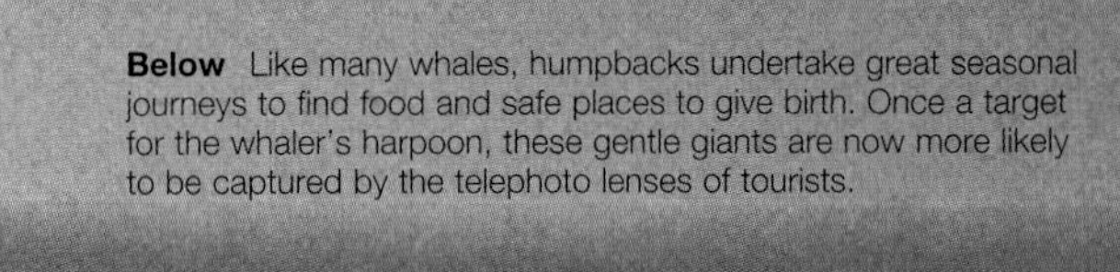

Below Like many whales, humpbacks undertake great seasonal journeys to find food and safe places to give birth. Once a target for the whaler's harpoon, these gentle giants are now more likely to be captured by the telephoto lenses of tourists.

MIGRATION BY AIR

6

50

36

33

Life on the Move

At any moment somewhere in the world millions of migratory animals are on the move. From fleet-footed antelopes to colossal whales and featherweight butterflies, an extraordinary variety of species embark on long and difficult journeys across land, through rivers and oceans, and in the air.

Migration is at once complex and mysterious. How do animals manage to travel so far and with such navigational accuracy? What is the strange gravity that their ultimate destination seems to exert on them? No wonder the phenomenon has captivated us for thousands of years, ever since Palaeolithic hunter-gatherers learned to follow herds of hoofed mammals across the grassy plains of what is now Africa and southern Europe.

Migratory animals have long been powerful symbols of change and renewal in human cultures. These seasonal events serve as a reminder that, to quote poet Ted Hughes (1930–1998), "the globe's still working". However, it is only in the last 150 years, and the past few decades in particular, that zoologists have really begun to reveal the hidden secrets of this fascinating animal behaviour.

Rapid developments in consumer electronics and mobile communication are today driving a revolution in migration studies. Satellite telemetry, which enables researchers to pinpoint the location of an animal tagged with a radio transmitter, is now so advanced that it is practical to follow the movements of individual creatures, almost anywhere on the Earth's surface, virtually as soon as they happen.

Left Flocks of migrating songbirds appear like green continents on this radar image captured at night. Radar is one of the many tools used to study migration.

The sooty shearwater, a globetrotting seabird, has been tracked across the entire Pacific Ocean as it traced a huge figure-of-eight pattern on a 64,000-km (39,800-mile) round trip from its nesting site in New Zealand. Beneath the waves, a leatherback turtle was monitored for 21 months on the longest recorded migration of any aquatic species, while a great white shark tagged off the coast of South Africa proceeded to swim over 20,000 km (12,400 miles) to Australia and back again in less than nine months. Data like these are helping scientists to piece together hitherto poorly understood animal journeys.

INCREDIBLE FEATS

Migration exists in a plethora of forms, from restive armies of caribou marching across the Arctic to the lonely odyssey of a tiny hummingbird over the Gulf of Mexico. But if one thing unites all migratory animals, it is the fight to survive. Migration is not nearly as dangerous as one might expect – it is a means of staying alive, after all. The participants have evolved sophisticated ways of reducing the risks so that, although some undoubtedly will perish, as many complete the trip as possible.

This book is a celebration of the greatest travellers in the animal kingdom. It also suggests how to experience some of them at first hand. Properly managed, ecotourism can make an important contribution to the struggle to save threatened species, so your own pilgrimages may play a small part in their conservation. There is no need to visit famous migration hotspots, though. Migratory wildlife can usually be enjoyed close to home – even in your garden.

TOP 10 MIGRATION RECORDS

Largest migrant	Blue whale	24–27 m (80–88 feet) long
Smallest migrant	Copepod (marine crustacean)	1–2 mm (0.04–0.08 in) long
Fastest migrant	Common eider duck	Average speed in still air: 75 kph (47 mph)
Rarest migrant	Amsterdam albatross	World population: 70–80 adults
Longest mammal migration	Humpback whale	Up to 8,500 km (5,250 miles) each way
Longest insect migration	Monarch butterfly	Up to 4,750 km (3,000 miles) in the autumn
Longest migration on foot	Caribou	Up to 6,000 km (3,700 miles) a year
Highest migration	Bar-headed goose	Maximum altitude: 9,000 m (29,500 feet)
Longest recorded round-trip	Sooty shearwater (tagged in New Zealand in 2005)	64,037 km (39,790 miles) across Pacific Ocean in 262 days
Longest recorded journey in water	Leatherback turtle (tagged in Indonesia in 2003)	20,558 km (12,774 miles) across Pacific Ocean in 647 days

Above The flights of common cranes that grace European skies each spring and autumn have entranced human observers since time immemorial. In the *Iliad*, Homer compared their resonant contact calls to the sound of approaching armies.

How Migration Works

A huge research effort is devoted to migration, yet our grasp of this endlessly fascinating subject remains patchy at best. Part of the problem is that migration takes many forms. It is much more than a simple trip from A to B, and migratory journeys are as varied as the animals that perform them. Some of the most intriguing questions in migration studies are how animals get ready, know when to set off and where to go, and navigate without getting lost. By monitoring the changing fortunes of migrant species, we can learn much about the health of the environment.

What is Migration?

Migration might be one of the great wonders of the natural world, but as a biological concept it is surprisingly fluid and elusive. Even today, there is no universally accepted definition. Animals make all kinds of different movements – short and long, seasonal and daily, regular and once-in-a-lifetime, highly predictable and seemingly random. It is not always easy to decide which ones are true migrations.

The classic idea of migration – and certainly the most widely held – is of flocks of birds flying north and south between separate breeding and non-breeding ranges in tune with the ebb and flow of the passing seasons, or pods of whales travelling to faraway feeding or birthing grounds. Many species from diverse animal groups do conform to this general migratory pattern, but it represents only one type of migration. There are numerous others, including journeys between east and west, complex circuits of land and ocean, seasonal trips up and down mountains, and vertical movements through the water column of seas and lakes. In addition, members of a particular species may follow a wide variety of migratory routes, and in some migrations only a portion of a species' population may be involved.

Below The common poorwill, a nocturnal species of bird found in western North America, is one of only a few birds to hibernate. Part of its population heads south to warmer climes in winter, while the remainder hibernates in nooks and crannies in rock.

So, migration has a wealth of meanings. However, for the purposes of this book, it is categorized as a journey with a clear purpose from one area or region to another, often following a well-defined route to a familiar destination, and often at a specific season or time. Any living thing that takes part in a migration is known as a migrant. Those that do not are said to be resident or sedentary.

WHY MIGRATE?

Put simply, migration is crucial for survival. It has evolved to enable animals to spend their life in two or more different areas, usually because a lack of food or a period of extreme weather makes it impossible to remain in the same location permanently. Other common reasons to migrate include: to find water or essential minerals; to hunt for a mate; to give birth, lay eggs, or raise young in a safe place; and to avoid predators or troublesome insect parasites. Animal migrations may be driven by several factors simultaneously.

ALTERNATIVE STRATEGIES

If an animal is to remain *in situ* when conditions become less favourable, it has three main alternatives to migration, at least in theory. First, it could adopt behavioural changes, such as in its diet or the shelter it uses. Second, it could undergo morphological (bodily) changes, by growing thicker fur or plumage for instance. Third, it could fall into a deep sleep, called torpor or hibernation – this is how many rodents, bats, bears, frogs, and toads cope with inhospitable winters. Amphibians and reptiles endure droughts in a similar state of dormancy, known as aestivation. Insects, too, are able to enter a kind of suspended animation, the diapause, to live through spells of harsh weather. In practice, however, such strategies are not a realistic proposition for large numbers of animals, which forces them to migrate instead.

ALMOST MIGRATIONS

It is probably best to regard some routine animal journeys as "almost" migrations rather than migrations in the strict sense of the word. For example, during the breeding season many parent animals temporarily leave their young to go on feeding trips. These sorties last up to a day in the case of seabirds such as boobies and gannets, or about three to five days in seals and walruses. Land carnivores, including wolves and spotted hyenas, are also long-distance commuters while they have offspring to provide for, frequently travelling dozens of miles to bring back fresh meat for their hungry cubs.

Below The life of an Antarctic fur seal is punctuated by frequent trips to catch fish, but brief feeding sorties like these, however regular, should not be classed as true migrations.

NON-ANIMAL MIGRANTS

Sometimes it is argued that there are migrants besides animals. Plants have the chance to extend their boundaries at the seed or spore stage of their life history, using the wind, water, and animals as agents for travel, and in certain respects, humans can be considered migrants as well.

The migration of *Homo sapiens* to every corner of the planet is relatively recent in evolutionary terms, having taken place within the last 50,000 years. In modern times, the most dramatic and far-reaching instance of human migration is emigration, such as that triggered by the Irish Famine of the 1840s. By the end of 1854, nearly two million people – about a quarter of the Irish population – had sailed west to the United States in search of a better life.

Below top Ladybirds survive temperate winters by huddling in clusters, but a surprising number of insects are strong migrants and fly to warmer regions when the cold hits.
Below bottom Humans are migrants, too. This illustration, published in the *Weekly Herald* in 1845, depicts newly arrived Irish emigrants in New York. The flood of refugees that fled from Ireland to the USA in the 19th century had a powerful and lasting impact on the histories of both countries.

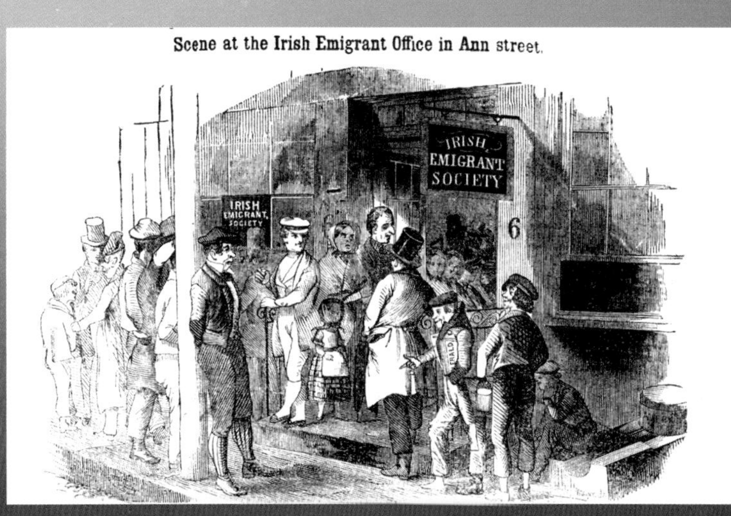

The Cycle of the Seasons

The Earth's movement and orientation in relation to the Sun govern the planet's seasonal cycles, which have a profound impact on wildlife. The changing seasons drive the migration of many species. Areas that provide a good living in summer become hostile in winter, pushing whole communities of animals to find more amenable conditions elsewhere.

An important feature of the global climatic system is the dramatic shift in climate according to latitude. Areas near the poles (high latitudes) have long, dark, bitterly cold winters followed by short, intense summers, whereas equatorial areas (low latitudes) enjoy high temperatures and a steady supply of sunshine all year round. While the Northern Hemisphere basks in summer, the Southern Hemisphere is locked in winter gloom and vice versa. These stark contrasts are caused by the fact that the Earth's rotational axis, on which it spins once each day, is tilted 23.5° from the vertical with respect to the plane of its orbit around the Sun. This results in one pole being tilted towards the Sun and the other away from it for six months each year; during the other six months, the position is reversed. The seasonal cycle produces huge differences in the duration and intensity of solar energy received in each hemisphere at any given time and has caused many species to evolve a migratory lifestyle.

LATITUDINAL MIGRATION

Many birds that breed in the Northern Hemisphere fly south for the winter, especially those that rely on insect food. They include a wealth of waders and waterfowl, together with vast numbers of passerines, or perching birds – warblers, flycatchers, thrushes, chats, shrikes, larks, pipits, finches, buntings, swallows, and martins. Birds of prey, storks, and cranes head south, too. To take Europe as an example: approximately 215 bird species migrate to sub-Saharan Africa after the breeding season, as part of a late-summer exodus that involves an estimated 5,000 million birds. On the other side of the globe, more than 300 species of bird make the trip from North to South America. Likewise, many birds from northern Asia spend the winter in the Indian subcontinent or the tropical southeast; a few, such as spine-tailed swifts, even get as far as Australia by island-hopping through the Indonesian archipelago.

Bird migration in the Southern Hemisphere is, broadly speaking, a mirror image of that in the north. Species here often fly north for the austral (southern) winter and south again for the austral summer. For example, vermilion flycatchers frequent the tropical savannas of Brazil in winter, staying there from March until August, then fly down to the fertile pampas of Argentina and Uruguay, where they rear their young between September and February.

Seasonal north–south movements are commonest in birds, although they do occur in other animal groups, particularly mammals (caribou, polar bears, and certain bats) and insects (various butterflies, moths, and dragonflies). In oceans, the best-known latitudinal migrants are whales, seals, and walruses. Whales, despite their status as long-haul travellers, seldom cross the equator. Instead, they remain in one hemisphere, passing the summer in high latitudes and retreating to lower latitudes for the winter.

Left The constant realignment of the Earth and the Moon relative to each other and to the Sun causes the seasons and tides, and these in turn have given rise to the phenomenon of animal migration.

TIME FOR A CHANGE

One of the most unusual types of migration is a seasonal excursion to a traditional moulting place. Ducks and geese briefly become flightless when they replace their wing feathers in late summer, so to reduce the risk of being caught by predators, they first fly to a secure, food-rich area, such as a shallow lake or sea, and moult on arrival. Uniquely among whales and dolphins, beluga whales also go through a concentrated period of moult in summer. Large groups of beluga whales enter shallow Arctic waters and rub themselves against the bottom to remove the old layer of outer skin. Often they swim up estuaries into fresh water, probably because this speeds the moult process. Perhaps the strangest moult migration belongs to yellow-lipped sea kraits of the tropical Indo-Pacific. These sea snakes periodically return to land to slough their skin, and marked individuals have been shown to navigate to the same beach every time.

Right Yellow-lipped sea kraits faithfully return to their "home" beach to shed their skin, even after feeding in deep water many miles from the coast. They also breed on land.

ALTITUDINAL MIGRATION

Climbing a mountain is comparable to making a much longer journey towards the poles, because the air cools with altitude by about 6.5°C per km (3.5°F per 1,000 feet). The decreasing temperature creates a succession of microclimates, each with characteristic habitats, which montane species use to their advantage. They migrate upslope in spring, and then downslope in the autumn. This is known as altitudinal, or vertical, migration.

Mountain-dwelling birds and mammals tend not to relocate in a single hop, preferring to move gradually, and halt whenever food is available. In temperate regions, their staggered descent is determined by the advancing snow coverage or falling temperatures, and their ascent tracks the spring thaw; in the tropics, montane birds travel in response to seasonal fluctuations in the supply of fruit and nectar. Mountain quails, native to shrubby uplands in westernmost North America, are among several gamebirds that migrate on foot. The quails walk in family groups up to 20 strong, and in central California will descend 1,500 m (5,000 feet) in winter.

Unlike latitudinal migration, which requires an ability to orientate along a fixed bearing with reasonable accuracy, altitudinal movements can be in any direction that reaches lower ground. Seasonal altitudinal migrations take place worldwide, but are a hallmark of the largest mountain chains.

Below Blue whales are fast swimmers that undertake epic seasonal journeys. However, in common with other whales, they do not traverse the globe as some birds do – rather, they stay in either the northern or southern half of the world.

Above In the tropical forests of Southeast Asia, male diamond pythons carry out a form of migration whose sole purpose is to locate and impregnate receptive females. After their reproductive odyssey, the snakes return to their solitary existence.

The Urge to Breed

There comes a time when animals must find a mate, and hunt for a safe place for their precious eggs or offspring to develop. Reproduction is a critical stage in the life history of any living creature, one which compels many species to make a special migration. In some cases, it will be their final journey.

Mammals often have a social structure based on sexual segregation, where the adult males live apart, in bachelor groups or as loners. This system, which is typical of herbivores, forces males in breeding condition to search for females in heat. Males may converge on traditional courtship grounds – for example, deer, antelope, and wild sheep assemble at the annual rut. Alternatively, in species such as elephants and rhinos, each adult male embarks on a solitary, testosterone-powered expedition to locate females. There is a little-known equivalent in snakes. Male diamond pythons from Southeast Asia use their acute sense of smell to track down receptive partners, and a "ripe" female python is likely to be pursued by an excited retinue of males for two or three weeks.

RETURN TO WATER

Due to a quirk of evolution, amphibians are the only terrestrial vertebrates alive today that develop from a larval stage. Their larvae are aquatic, equipped with water-breathing gills instead of lungs. Inevitably, therefore, the adult phase of an amphibian's life-cycle is punctuated by a series of return migrations to fresh water to breed. Numerous frogs, toads, and salamanders carry out spawning migrations, which vary enormously in both length and choice of breeding habitat. For some species, any nearby puddle will suffice, whereas the rest may trek to a pond or marsh up to a few miles away, and revisit the same locality year after year. Amphibians migrate on wet nights, to keep their permeable skin damp. The cue for their departure is the first cloudburst of the rainy season or a sharp rise in temperature in spring.

MIGRATING GENES

Migratory behaviour is seldom seen in primates. Most inhabit tropical forests, where there is a stable climate and year-round food supply, and form mixed-sex groups, so potential mates are always available. However, female chimpanzees shift from one community to another, moving as far as 15 km (9 miles) on occasion. Studies of chimp DNA suggest that in this way female genes can "migrate" several hundred miles over successive generations.

RETURN TO LAND

Among reptiles, the principal type of migration is to and from traditional egg-laying sites, undertaken by freshwater and sea turtles. Like their ancestors, these turtles produce soft-shelled eggs with the texture of parchment, which obliges them to lay on dry land. They need beaches with just the right gradient and sand, hauling ashore to dig pits that act as efficient incubators. Since suitable beaches are few and far between, the turtles tend to nest communally in the best spots. Similar behaviour is found in various iguanas from the Galapagos Islands and Central America, which migrate to a handful of nesting areas.

Reptiles are not alone in making repeat migrations to long-established nest sites. The pinnipeds (seals, sealions, and walruses) have retained their ancestral link with the land: unlike whales and dolphins, they have not solved the problem of giving birth at sea. Sheltered beaches, rocky coasts, and ice floes serve as their nurseries. Seabirds cannot rear their young at sea, either; consequently, they flock to coastal cliffs and remote oceanic

Above On a few nights each June and July, the strandline of beaches in Newfoundland is carpeted with silvery capelin. The wriggling masses of 15-cm (6-inch) fish arrive suddenly on the high tide to spawn, then depart equally quickly.

islands to nest in their tens of thousands. The breeding migrations of shearwaters, albatrosses, and terns are some of the longest and most spectacular in the entire animal kingdom.

RULED BY TIDES

The Moon is unusually large and close compared to the moons of other planets, and so has a powerful influence on life on Earth. Extreme forms of human behaviour are reputed to happen around the full moon – hence the word "lunacy" – but, by virtue of the tides that it causes, the Moon's impact on marine animals is greater still. Marine fish, turtles, and invertebrates all synchronize their reproductive migrations with the tides, often choosing a particular moment in the lunar cycle.

Huge numbers of American horseshoe crabs, distant relatives of spiders, swarm across beaches at full moon during high spring tides. Timing their spawning to coincide with the year's highest tides allows these prehistoric-looking arthropods to ensure their eggs are deposited on the upper zone of the beach, comfortably beyond the reach of scavenging intertidal creatures. Other tide-driven breeders include two small fish – California grunion and capelin – which surf in on the waves en masse and fling themselves onto the wet sand to mate and lay their fertilized eggs.

BODY SHOCK

Some fish divide their lives between fresh and salt water, using one aquatic world as a nursery and the other as a spawning ground. When moving from river to sea, or vice versa, they have to withstand a huge change in the concentration of salts and other minerals in the water. The fish have specially adapted kidneys and gills that enable them to reverse their osmotic functioning. Without these adaptations, they would lose water upon entering the sea, or swell up like a balloon when travelling in the opposite direction. Fish that migrate downriver to the sea to spawn, such as American and European eels, are called catadromous; species that head upriver, including salmon, sea trout, sturgeon, and shad, are described as anadromous. Most spawn once and die soon afterwards, but the beluga, or European sturgeon, might make the journey dozens of times and live for over a century.

Below Common frogs mass at their breeding pond. Though amphibians do not migrate far, due to the risk of exposure, they often make the journey in impressive numbers.

Nomads and Invaders

Not all animal journeys are in response to seasons – some lack a fixed destination or route. Wholly unpredictable, these movements include itinerant wandering, large-scale invasions, and spontaneous flights to escape bad weather or volcanic eruptions. Sometimes pioneering individuals encounter ideal living conditions and remain in their new territory as colonists.

The ability to be flexible in movement patterns is an important aspect of coping with environments that have a patchy food supply or irregular rainfall. It is more beneficial to move from place to place according to need than to follow a pre-programmed migratory regime. Such a lifestyle is known as nomadism and is characteristic of species that live in grassland and desert habitats. The savannas of the African tropics, for example, are home to legions of highly mobile hoofed mammals, which hop between the fertile green "islands" produced by localized rain. Their temperate counterparts are skittish herds of Mongolian gazelles and saiga antelopes, which have adapted to survive in the parched grassy steppe that covers vast swathes of Central Asia, where standing water is almost non-existent.

Interestingly, birds that undertake long latitudinal migrations are often sedentary in their nesting areas and nomadic while on their wintering grounds. For example, North American wood warblers join

Facing page Saigas spend their entire lives roaming the arid steppe of Central Asia. Intense hunting has led to a collapse in their populations. **Above** Some invasive species can be highly destructive, and none more so than locusts. Here, a Filipino farmer tries in vain to protect his fields from a swarm of Oriental migratory locusts.

Above Swifts are like living barometers. They are specialized aerial feeders, so as soon as the atmospheric pressure falls and the weather takes a turn for the worse, they escape by flying to unaffected areas. Swift nestlings can go without food for several days while their parents are away, by living off their fat reserves.

THE ARID CONTINENT

Australia is the driest inhabited continent: two-thirds of its surface is desert or semi-arid bush – a landscape called the "outback". In the absence of a clearly defined seasonal cycle, wildlife can survive here only if it is able to handle prolonged drought and widely fluctuating conditions. For this reason, the Australian mainland has the greatest diversity of nomadic birds on the planet. One in three of all its breeding birds, from budgerigars to emus, are nomads. Kangaroos and wallabies are among the outback mammals that lead a nomadic existence.

roving flocks of local birds in the forests of Central and South America, and Arctic-breeding thrushes spend the winter roaming the agricultural land across Europe and Asia's middle latitudes.

INVASIONS

An "invasion" or "irruption" is when a more or less sedentary population of animals is suddenly impelled to move as a result of overcrowding or food shortages. From an ecological perspective, it takes place at the moment when living standards have declined so far that the situation becomes untenable and mass emigration is now the best option. Irruptions are erratic events that drive species beyond their usual range, and occur in a variety of birds and rodents from the far north; most famously, in lemmings. They are an important phenomenon in the life-cycles of many insect pests, including locusts, planthoppers, and cutworms (the crop-munching larvae of certain moths). Forecasting when the next outbreak might be due is an imprecise science, but one that has grown into a multimillion pound enterprise.

ESCAPE MOVEMENTS

Some birds react very quickly to deteriorating weather, which forces them to move elsewhere without warning; the human equivalent would be sunbathers dashing for cover during a squall. This type of weather-induced exodus, called an escape movement, can mean the difference between life and death to a bird. The most dramatic escape movements are carried out by swifts – superbly aerobatic species with scythe-shaped wings, which trawl the sky for flying insects. Swifts are supreme fair-weather birds, totally reliant on clear skies to find food. At the first hint of an approaching low-pressure system, they circumnavigate the depression, flying over or under the front to reach the calmer air behind it. This may involve a round-trip of over 2,000 km (1,250 miles) in only a few days.

YOUTHFUL STRAYS

Young animals often appear to be hard-wired with a powerful wanderlust. The exploratory urge takes them away from their birth or hatching site, so that they can familiarize themselves with their neighbourhood. Additional benefits of juvenile dispersal are that it helps a species to spread out through all of the available habitat, and reduces the risk of inbreeding.

Juvenile dispersal is seen in many animal groups. The chicks of some seabirds, such as guillemots and razorbills, leave their nests before they are able to fly, and so, accompanied by a parent, they leave their coastal breeding colony by swimming out to sea. Similar behaviour is found in other members of the auk family, Alcidae, and also in penguins. Juvenile king penguins may swim over 1,000 km (600 miles) from their natal colony, while young African penguins, which start life on the coasts of South Africa, frequently reach the Atlantic's equatorial waters.

Keeping Time

In essence migration is about being in the right place at the right time, and so migratory species need to have some form of inbuilt clock. Accurate time-keeping allows an animal to keep in step with changes in the outside world, and to begin and end journeys on cue. It is also essential for successful navigation.

Above This black-tailed godwit is performing its courtship flight in spring. The species has developed a system for synchronizing its migrations so that both male and female arrive at their breeding grounds at the same time.

The natural world exhibits many astonishing feats of time-keeping: sea turtles and crabs return to their nesting beaches on the same few nights each year, shoals of fish appear at certain locations in the ocean on a predictable schedule, and generations of migratory birds arrive back on their breeding grounds within a couple of weeks of their species' traditional date. Punctuality clearly is a widespread phenomenon in animals. In fact, we now know that virtually all living things, from bread mould and fruit flies to human beings, possess internal time-keeping mechanisms. In 2008, scientists announced that they had discovered "clocks" inside individual human cells, and it seems fair to assume that such mechanisms exist at a cellular level in animals as well.

ARRIVING TOGETHER

Migratory birds that pair for more than one nesting season face a dilemma – if partners spend the winter apart and travel to and from their wintering areas separately, how can they ensure that they return to their breeding ground at the same time? If either bird were to turn up early, it might find a new mate before its old partner eventually arrived, leading to "divorce". Paired migrants can avert this possibility due to an amazing capacity to synchronize their migrations. Ornithologists monitored the arrival dates of black-tailed godwits at their nest sites in Iceland in spring, and found that established pairs of these large waders showed up within three days of each other, despite being separated for months and flying thousands of miles. It is a mystery how the godwits time their reunion with such extraordinary accuracy.

DAILY AND ANNUAL RHYTHMS

So what are internal clocks exactly? The answer is complex, and the way they work is not yet fully understood. Some types of animal behaviour, such as self-defence, can be provoked spontaneously; they do not operate according to a timetable. However, lots of other processes, including feeding, sleeping, metabolism, and reproduction, are governed by strict 24-hour cycles. These cyclical patterns of activity are called circadian rhythms, from the Latin *circa*, meaning "around", and *diem* or *dies*, meaning "day". They may be influenced by external cues – changing temperature or humidity, the rise and fall of the tide, or the alternation between night and day, for example – but are internally generated and instinctive. Circadian rhythms explain why people feel hungry at lunchtime or sleepy in midafternoon (siesta time), why shiftworkers have difficulty adapting to a new rota, and why we suffer from jet lag when crossing many time zones in less than 24 hours.

In addition, there are long-term regulatory cycles, known as circannual rhythms. As before, while these respond to stimuli outside the body, such as gradually changing day length and the yearly sequence of passing seasons, they are primarily internal. Circannual rhythms are most developed in species found in the world's temperate regions, especially near to the poles, where the annual change in day length is greatest.

Circadian and circannual rhythms work together to create perfectly calibrated, extremely efficient clocks. They are unique to each species, having evolved to suit its particular lifestyle and environment. It follows that these deep-seated mechanisms are fundamental to the ability of migratory species to plan and coordinate their journeys successfully. Prior to setting off, many of them become visibly unsettled. The pent-up excitement is most noticeable in birds, which flutter their wings and change perch frequently, an agitated behaviour known by the German word *zugunruhe*.

Above The changing level of melatonin in the bloodstream helps animals such as this African elephant to keep time and coordinate daily activity. Production of the hormone peaks at night, then plunges during the daylight hours.

PACEMAKERS

The majority of animals have a "pacemaker" in overall control of their circadian and circannual rhythms. In mammals, the pacemaker is a part of the brain known as the SCN, short for suprachiasmatic nucleus. The hormone melatonin also has a vital role to play. Melatonin is secreted by the pineal gland, but only at night, because daylight blocks its production. Therefore, day length determines how much of it the body makes, and the melatonin level helps to regulate daily and seasonal activity. The pineal gland appears to act as the master pacemaker in fish, reptiles, and amphibians.

Sometimes migratory animals have to suspend their normal circadian rhythms temporarily. For example, species that travel to the Arctic or Antarctic for the polar summer encounter near-constant daylight, which means their circadian rhythms based on the daily light–dark cycle cannot function. Animals such as caribou therefore become "non-rhythmic" during the polar midsummer.

Above Birds often assemble at precise times of the day and year. Here, European starlings perform an aerial ballet before settling down to roost, as they do every winter evening.

TIME AND NAVIGATION

Migrants that orientate using visual clues, such as the position of the Sun and stars, have to allow for the Earth's rotation, which varies but is approximately 1,700 kph (1,050 mph) at the equator. If a trans-equatorial migratory bird was just five minutes out in its timing, the error would result in a diversion of almost 160 km (100 miles). For centuries this problem plagued human navigation at sea, until the development of accurate portable clocks in the 1750s enabled sailors to find their precise longitude.

Surviving the Journey

Migration is unforgiving on those who take part. It can be a relentless struggle and often puts animals under enormous stress, pushing their metabolism and other body processes to the limit. The odds might seem to be stacked heavily against ever reaching the destination. But the reality is that, generally, most migrants do arrive unscathed, thanks to a suite of physical and behavioural adaptations.

Some species are spectacularly ill-suited to a migratory life. Big cats for instance do not make good migrants, since they produce young that are helpless for many months. Size can be an important factor, too: the majority of small land animals simply cannot afford the energetic cost of migration. Most rodents lack the endurance to cope with regular long-distance travel; a 100-g (3.5-oz) rodent would use about 25 times more energy per unit of body mass than a 200-kg (450-pound) antelope.

By contrast, the entire life history of other animals may be geared towards migration. Antelope and gazelle calves are on their feet only minutes after birth, and, due to their disproportionately long legs and the extremely high fat content of their mothers' milk, they can keep up with the rest of the herd within days. The young of tundra-nesting waders develop so rapidly that they can embark on their maiden southbound migration from the Arctic when only two months old. And sea turtle hatchlings are already proficient swimmers and head straight to the safety of deep water.

ADVANCE PREPARATION

Of course, not all migrants begin to migrate when still babies – many have to ensure that they are in a fit state first. In adult birds, this preparation includes a moult, the timing of which is controlled by hormones and the birds' inbuilt circannual rhythm (see pages 18–19). Replacing old and worn plumage is crucial, because flight efficiency depends on the condition of the wing feathers.

Migratory animals often feed intensively prior to departure.The aim of this gluttonous behaviour, known as hyperphagia, is to boost fat reserves to use as fuel. Hyperphagia is switched on automatically by an internal circannual rhythm and is seen in animals as varied as monarch butterflies, caribou, and baleen whales; in insects, it can lead to a 30 per cent increase in body weight, and whales sometimes double their normal weight. But the migrants-to-be do not only eat more – they also look for particularly high-energy foods. In temperate latitudes, insectivorous birds such as warblers and thrushes shift to a sugary diet of fruit in late summer and autumn to lay down a thick layer of fat before they set off.

Finally, animals may go through a radical physical transformation. Birds develop larger, more powerful breast muscles and shrink non-essential organs accordingly (to avoid excessive weight, which would hamper efficient flight). Some insects do much the same – for example, the generation of monarch butterflies migrating south through North America in the autumn has no sexual organs – these develop the following spring. And desert locusts grow longer wings and look like a completely different creature by the time they take off.

Below Whooper swans, which weigh 8.5–10 kg (19–22 pounds), are about the maximum size for long-distance avian migration. Large birds cannot lay down extra layers of fat in the way that smaller species do, because they are already close to the weight limit that their wings can bear.

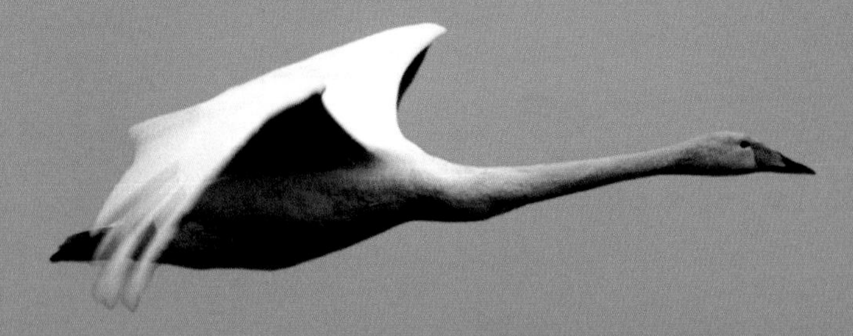

Left For safety, Atlantic herring migrate in huge, tightly packed shoals, which sometimes contain millions of fish. The largest shoals on record filled a volume roughly equivalent to 4 million litres (1 million gallons) of water.

MANAGING THE RISK

Over thousands of years migratory animals have evolved many solutions to the problem of how to make their journeys less dangerous. To combat the ever-present threat of predators, migrants frequently travel in groups or at certain times of day or night. Obviously it is a good idea to reduce the effort involved by taking advantage of favourable environmental forces, such as the wind or ocean currents (*see* pages 22–23), and by finding the right pace. There are both fast and slow migrants – every species follows its own migratory timetable suited to its strength, stamina, and fat reserves, and the distance to be covered.

Staggered migration, with stops to rest and refuel, is another common strategy. Bats visit a series of conveniently located roosts during their journey, while swarms of butterflies and moths settle on trees and buildings to roost overnight or until a spell of bad weather has passed. Traditional locations for resting are called stopovers, or staging areas. The best are used year after year, and at key periods host vast gatherings, particularly of birds. For example, approximately 45 per cent of all migratory waders nesting in North America stop to rest, or "stage", at Cheyenne Bottoms, Kansas, in spring. Protecting major staging areas such as Cheyenne Bottoms is therefore a conservation priority.

LIMITS OF ENDURANCE

Some birds are high-revving migrants that push themselves to their physical limits, leaving little "gas in the tank". In autumn, the Icelandic population of whooper swans flies over the North Atlantic to wintering areas in Britain. The rapid non-stop sea crossing to western Scotland takes 12–13 hours at average speeds of 65–80 kph (40–50 mph), which is extremely close to the physical limits for such large, heavy birds.

HALF A MIGRATION

Surely the world's strangest migrations are those where only half of the animal survives the journey. Various marine worms in the class Polychaeta migrate in this bizarre fashion, including North Atlantic ragworms and palolo worms from South Pacific reefs. Each worm splits in two and the tail end of the animal, called the epitoke, swims away using paddle-shaped lobes on each of its segments. At night the tail rises to the surface of the ocean together with incredible numbers of other worm tails. The swarming tails then burst open to shed eggs and sperm into the water, which fuse to produce larvae, thus restarting the worm's life-cycle. Swarming is usually synchronized with the moon cycle.

Above Sardines use a powerful cold current to migrate up the coast of South Africa's Eastern Cape province, where they provide food for dolphins and other fish eaters. The predators round up the fish into bait balls.

A Helping Hand

The forces of nature offer welcome assistance to tired migrants. Insects and land birds ride tailwinds and circle upwards in thermals, seabirds catch the updrafts created by waves, and turtles and fish are pulled along by ocean currents. As a result, the Earth's prevailing winds and currents exert a powerful influence on the timing and direction of migratory journeys.

THE SARDINE RUN

Each year in June or July, a finger of cold water advances up the coast of southeast Africa towards Madagascar. This is the starting signal for the annual Sardine Run, in which huge shoals of sardines leave their usual cold-water haunts further south, and push north into subtropical areas that are normally too warm for them. Some of these shoals are up to 6.5 km (4 miles) long. A flotilla of hungry predators follows in their wake, including common and bottlenose dolphins, yellowfin tuna, king mackerel, and several species of shark. When sea temperatures rise above 21°C (70°F) again, the sardines return south and the feeding frenzy comes to an end.

The oceans are never placid like lakes, even on the most halcyon days. Although the sea may look calm at certain times, currents and upwellings surge below the surface. In addition, the tidal cycle has a profound impact on marine life, affecting water circulation far from land, especially during the largest "spring" tides that occur twice each month. So it comes as no surprise that pelagic species – inhabitants of the open ocean – should make full use of their constantly moving three-dimensional habitat, by ascending and descending through the water column to find a powerful flow that will propel them in the right direction. There is an added bonus to swimming in fast currents: they are often laden with food.

Many currents follow definite routes that can be drawn on a map (see right), and migratory animals develop a close relationship with them. In summer leatherback turtles pick up the warm Gulf Stream to glide across the North Atlantic to the coasts of northwest Europe, which teem with their favourite jellyfish prey, while the movements of manta rays and whale sharks show a clear correlation with warm currents throughout the tropics. The larvae of countless pelagic fish, molluscs, and crustaceans rely on currents to drift away from their spawning areas.

RESTLESS ATMOSPHERE

Just as the marine environment is in permanent motion, so too is the planet's air. Atmospheric conditions can change rapidly, which means birds and other aerial migrants have to time their departure carefully; sometimes a delay of only a few hours may be disastrous. The ideal scenario for migration is a sustained tailwind and cloudless skies. Lower than average temperatures are also a boon, because the cool air prevents hard-working pectoral muscles from overheating. (This is one reason why some birds migrate at night, particularly when flying across deserts where the punishing midday heat could prove fatal.)

Flying migrants avoid setting off on windless days, as the still air obliges them to use more precious energy. No birds are more dependent on the wind than albatrosses, most species of which use the wind to help them soar and glide over the tempestuous Southern Ocean on characteristically stiff wings (see Wave Riders, pages 122–123). Faced with a becalmed sea, these majestic ocean wanderers cannot stay airborne and are reduced to bobbing around on the surface like rubber ducks.

FREE RIDES

Two types of atmospheric turbulence have special importance for migrating birds: jet streams (see right) and thermals. When a bird enters a jet stream – a fast-flowing air current found at high altitude – it is able to cover enormous distances with astonishing speed. For example, a knot caught up in a jet stream might find itself moving at up to 240 kph (150 mph), which is considerably quicker than a diving peregrine falcon, ordinarily the fastest animal on the planet. Certain insects also take advantage of high-level winds, despite the fact that this strategy results in many individuals becoming fatally injured or getting blown off course. In 1976, a wind-assisted swarm of painted lady butterflies was spotted west of St Helena in the middle of the Atlantic Ocean, having been swept about 3,200 km (2,000 miles) from its probable point of departure in southwest Africa.

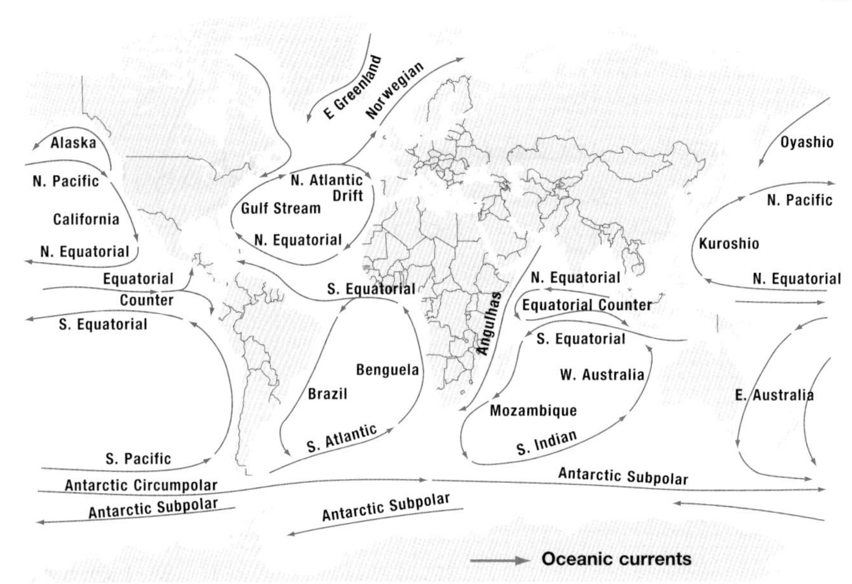

Above The main ocean currents have a major impact on marine migrations. Their position is determined by a variety of factors, including the prevailing wind, the Coriolis effect (produced by the Earth's rotation), the salinity and temperature of the water, and the topography of the seabed.

Thermals – rising bubbles of air warmed by the ground below – are crucial to the migration of large soaring birds, from eagles and hawks to pelicans and storks. Having found a thermal, these species ascend in lazy spirals, seldom bothering to beat their wings. Once they arrive at the top of the upcurrent, they glide out and away, gradually losing altitude until they locate another thermal and can repeat the procedure. "Thermal hopping" is an amazingly efficient mode of travel, but since thermals do not form over water, at night, or in cold weather, it limits where and when soaring migrants are able to move.

Below Some migratory birds ride jet streams. These air currents are invisible, but often their location is indicated by ribbons of cloud at high altitude.

Different Routes

Most migrants are creatures of habit that have a tried and tested route map. Rarely is this a straightforward line drawn directly between two points. Migratory journeys are shaped by the physical geography of the land and ocean, so looping routes and diversions around major barriers are common, and the outward and return legs may be different.

Animal migration generally advances along a broad front, which can be hundreds of miles wide. The front consists of many separate parallel streams of migrants simultaneously following the same bearing. If the progress of every individual taking part in the migration was plotted on a map at regular intervals, the resulting pattern would therefore resemble a wave, sweeping forwards in a long line. A huge variety of animals, from hoofed grazers to waterfowl, small songbirds, bats, butterflies, dragonflies, and land crabs, migrate in this fashion.

However, some species use a much more restricted migratory artery, known as a narrow front. This style of migration is found mainly in large land birds, such as storks, cranes, and birds of prey. It is also typical of coast-hugging ocean migrants, including grey and right whales, which follow the shore, seldom straying from the shallow waters of the continental shelf.

LEADING LINES

The shortest route is not necessarily the easiest or safest. Certain physical features, called leading lines, encourage migrants to adopt a particular path, regardless of whether their journey is extended. Leading lines include rivers, streams, lake shores, valleys, mountains, and coasts. They are used by terrestrial animals great and small, and by a wide range of aerial migrants too. Often they are in themselves natural barriers to progress – for instance, the Rockies, Appalachians, and Andes present a massive hindrance to west–east migration, yet enormous numbers of migratory insects and birds funnel down their flanks in a north–south direction.

Oceans also possess leading lines, which are equally important to those on land, albeit less obvious. Beluga whales head north

This page Great shearwaters have a "loop" migration that traces a giant circle around the Atlantic, from their breeding colonies on the Falkland Islands and Tristan da Cunha to the waters off eastern North America, the European coasts, and south again.

FLYWAYS

Migratory birds are just as influenced by the changing contours of the Earth's land as are migrants that travel on foot. The migratory avenues in the sky produced by geography are called flyways. This maps plots the major flyways in North and South America, Europe, Asia, and Africa, each of which is used by birds of many different species during peak migration seasons.

to their summer feeding grounds in the Arctic by swimming up "leads" – narrow cracks in the sea ice that serve as convenient expressways. Below the surface of the world's oceans, turtles and fish follow submerged mountain chains and move along the steep drop-off on the seaward side of coral reefs. Sharks appear to be able to detect and follow "roads" of magnetism across the seafloor (see pages 92–93).

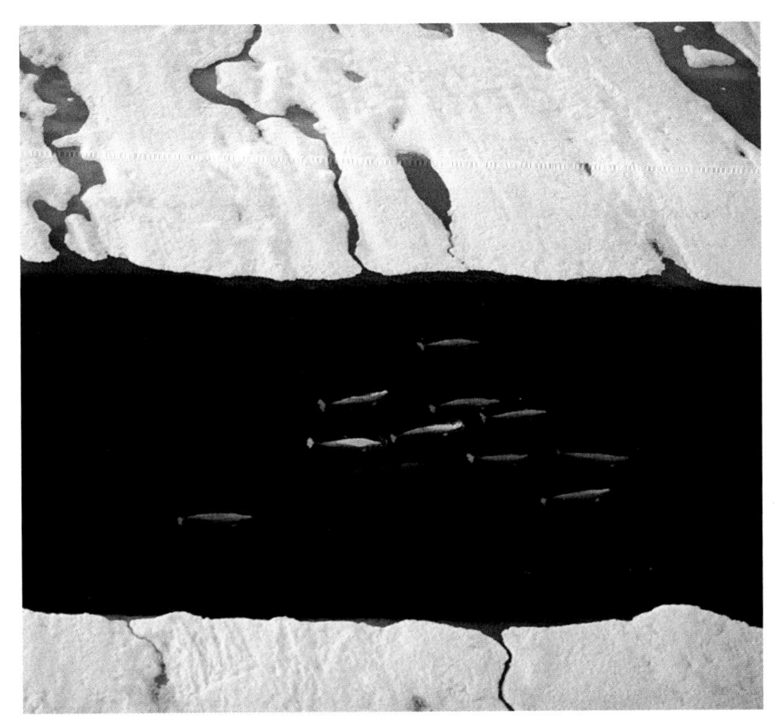

Above Migrating beluga whales follow traditional routes that may be hundreds of miles long, and readily make use of cracks in the Arctic's pack ice during their travels.

WAIFS AND STRAYS

Sometimes migratory journeys go badly awry and animals head completely off target. Birds, bats, and insects can all get into difficulties in low cloud and heavy rain, which severely hampers their ability to orientate. In the worst-case scenario they may end up far from their species' normal range, occasionally making landfall on ships and oil rigs in mid-ocean, or even on the wrong continent. Crosswinds are another potential problem: a gentle side breeze might seem harmless at first, but if it strengthens during the flight, an airborne migrant will unwittingly drift further and further off course.

In general, older, more experienced individuals are more likely to recognize and be able to compensate for these various difficulties. The vast majority of hopelessly lost migrants, which biologists refer to as vagrants, are juveniles on their first outing. Perhaps the most tragic case of migratory inexperience is when sea turtle hatchlings become confused by brightly illuminated beachfront developments, mistaking them for the dim glow over the ocean, with the result that the baby reptiles crawl inland instead of towards the breaking surf.

MIGRATORY DIVIDES

Different populations of a migratory species often have their own itineraries, even if the final destination is the same. This phenomenon is most evident in migrants that need to bypass a substantial obstacle, such as a desert or sea. When two populations take divergent routes like this, the split within the species is known as a migratory divide.

Visible Clues

Animals have evolved highly efficient direction-finding systems, in which visual clues often play a central role. Migrants look for familiar geographical features, orientate by the Sun, and interpret the movement of stars in the night sky, enabling them to steer the right course over great distances with pinpoint accuracy time after time.

Orientation – the ability to use various external clues to stay on the correct heading – is vital to migratory success. The other fundamental aspect to migration is navigation – the capacity to zero in on a particular location from somewhere else. If anything, the art of navigation is even more complex and mysterious, and no-one is quite sure how the different navigational systems used by migratory animals work together in practice.

LOOKING FOR LANDMARKS

Many migrants scan the landscape for short-range navigation, usually to find their precise target during the final stage of their journey. High-flying birds obtain a panoramic vista of the ground below, so look out for reference points such as rivers and coasts, building up a picture of their surroundings that they remember from one migration to the next. The horizon is in itself a helpful clue. Research has shown that homing pigeons seldom approach their loft along a straight compass bearing, but are guided by the twists and turns of major features in the area, including artificial ones such as roads and power lines. Even birds that migrate at night are helped by landmarks, from the glint of moonlight on water to the bright lights of a town.

Marine animals also are thought to navigate using knowledge of local topography. For example, seals and whales can probably recognize features of the seafloor. Some marine biologists have suggested that sea turtle hatchlings "imprint" on unique characteristics of their natal beach, and that years later adult females recall this information to help them locate the same stretch of sand.

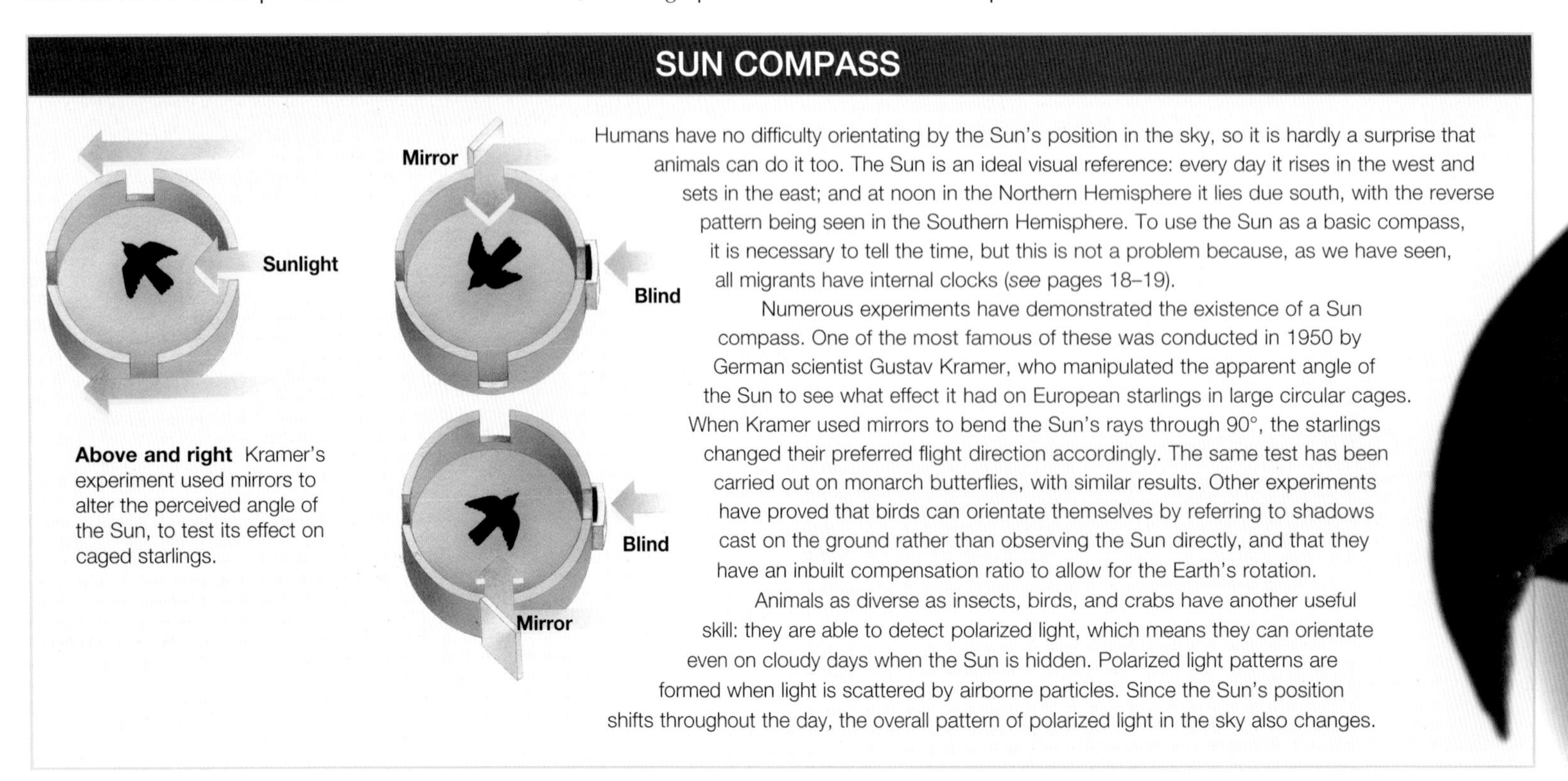

SUN COMPASS

Humans have no difficulty orientating by the Sun's position in the sky, so it is hardly a surprise that animals can do it too. The Sun is an ideal visual reference: every day it rises in the west and sets in the east; and at noon in the Northern Hemisphere it lies due south, with the reverse pattern being seen in the Southern Hemisphere. To use the Sun as a basic compass, it is necessary to tell the time, but this is not a problem because, as we have seen, all migrants have internal clocks (*see* pages 18–19).

Numerous experiments have demonstrated the existence of a Sun compass. One of the most famous of these was conducted in 1950 by German scientist Gustav Kramer, who manipulated the apparent angle of the Sun to see what effect it had on European starlings in large circular cages. When Kramer used mirrors to bend the Sun's rays through 90°, the starlings changed their preferred flight direction accordingly. The same test has been carried out on monarch butterflies, with similar results. Other experiments have proved that birds can orientate themselves by referring to shadows cast on the ground rather than observing the Sun directly, and that they have an inbuilt compensation ratio to allow for the Earth's rotation.

Animals as diverse as insects, birds, and crabs have another useful skill: they are able to detect polarized light, which means they can orientate even on cloudy days when the Sun is hidden. Polarized light patterns are formed when light is scattered by airborne particles. Since the Sun's position shifts throughout the day, the overall pattern of polarized light in the sky also changes.

Above and right Kramer's experiment used mirrors to alter the perceived angle of the Sun, to test its effect on caged starlings.

Facing page Birds are visual migrants par excellence, able to orientate by the Sun, stars, polarized light, and geographical landmarks. Their eyes have an internal structure called the pecten oculi, which is thought to cast shadows on the retina in the manner of a sundial. The bird pictured is a northern goshawk.

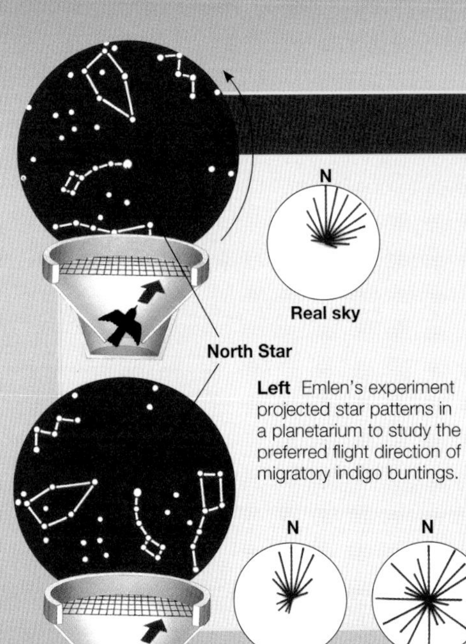

Left Emlen's experiment projected star patterns in a planetarium to study the preferred flight direction of migratory indigo buntings.

STAR COMPASS

Few birds that travel at night seem to orientate by the position of the Moon (in fact, moonshine is an annoying distraction). Instead, they observe changing star patterns. The crucial factor is the rotation of the night sky about a single fixed point, which in the Northern Hemisphere is the Pole Star. Nocturnal migrants have no need to see individual stars or constellations, so long as they are able to make out the centre of rotation.

The American biologist Stephen Emlen investigated the nature of the star compass using captive indigo buntings in a planetarium. The birds were placed on an ink pad at the bottom of a paper cone so that the resulting pattern of inky footprints indicated their preferred direction of travel. When the stars in the planetarium were obscured, the buntings became confused, hopping aimlessly. But when a real sky was projected, they soon located the Pole Star and used it to find north – the direction of their spring migration from Central to North America. When the entire sky was rotated, the birds still managed to head north, proving that movement of stars rather than their position is the key.

How bats orientate is less well known, although big brown bats and certain other species have been shown to use the lingering post-sunset glow to home in on their caves. The same technique might be employed by long-range migrants such as North American hoary and grey bats and European noctules.

Invisible Clues

Every migratory animal has a range of orientation mechanisms, many beyond the realm of human perception. Some species find their way by smell, taste, or sound. Others analyse subtle changes in water quality. Most remarkable of all is the ability to orientate by sensing tiny variations in the Earth's magnetic field.

The idea that animals can chart a course by sniffing seems incredible, but a wide variety of migrants do use their sense of smell for orientation. Wildebeest walk with heads bowed to follow a pheromone trail through the grass, laid down by scent glands in the hooves of individuals further up the line. Sea turtles approach their nesting beaches from downwind so that the distinctive aroma of land can guide them ashore. Even birds, which have a relatively poor sense of smell compared to other animals, orientate in this way: tests on homing pigeons demonstrated that they recognize the smell of their own neighbourhood; and it is thought that seafaring birds such as petrels literally follow their noses, heading towards the odour clouds released by ocean upwellings or the pungent whiff of their clifftop breeding colonies.

Aquatic animals, especially fish, have acutely sensitive olfactory organs that play a major part in orientation. Salmon migrating in the open ocean are able to locate the mouth of their natal river by constantly sampling the changing concentration of mineral salts in the seawater; they are following what is known as an olfactory gradient. Once in the river, they exploit the same technique to move upstream to their spawning grounds. Whales can probably distinguish the chemical signatures of different parts of the ocean and perhaps identify rich feeding areas by tasting the faint trace of distant plankton blooms.

Above Insect-eating bats, such as this little brown bat, probably use echolocation for navigation as well as for tracking down food.

SOUND

Migration is often a noisy affair. This is partly because groups of animals on the move need to communicate to stay together; another reason is thought to be that sound itself acts as an orientation system, at least over short distances. There are probably two main forms of orientation by sound. The simplest, called phonotaxis, is the ability to pinpoint sound sources. It is typical of frogs and toads, which are enticed to join others at the breeding pool by a chorus of croaks, burps, and chirps drifting through the night air. The second technique – echolocation – involves sending out pulses of sound and picking up the echoes to create a detailed picture of the surrounding area. The undisputed experts at echolocation are insectivorous bats (suborder Microchiroptera) and toothed whales (suborder Odontoceti), a group that includes sperm whales, dolphins, and porpoises.

Bats can detect only those objects within echoing distance – about 100 m (350 feet) – if this were their sole method of orientation they would need to remember an entire series of acoustic "signposts" lining the migration route. It is therefore likely that they back up echolocation with visual clues and a magnetic compass. By contrast, whale calls carry hundreds of miles underwater and may be used in long-range navigation: pods of sperm whales have been recorded repeatedly bouncing their powerful calls off the edge of the continental shelf, and one theory is that they were using sound to monitor their progress.

Ambient noise in the environment may also have a role in orientation. Since many birds and mammals can hear low-frequency sounds, which travel great distances, this raises the possibility that they listen for long-range acoustic clues such as waves on a beach or the wind whistling against mountains. However, the evidence is inconclusive.

MYSTERIOUS FORCES

Several geophysical forces besides magnetism may one day be shown to help animals stick to their migratory path. They include gravity; the Coriolis force, produced by the Earth as it spins; and barometric pressure, which is already known to inform birds about the best time to set off. However, scientists have yet to prove that any of these forces are used by animals to help them orientate themselves.

MAGNETISM

The Earth behaves like a giant magnet. It has a doughnut-shaped magnetic field, made up of elliptical force lines running between the magnetic north and south poles. The Earth's magnetism was long suspected to assist migratory animals, but the existence of a magnetic compass was not proved experimentally until the 1960s, when the German researchers Friedrich W. Merkel and Wolfgang Wiltschko placed European robins in cages surrounded by electrical coils and then subjected them to changes in the force field around them. If the field was altered so that magnetic north appeared to lie in a different direction, the robins modified their preferred flight path. Further tests showed that the robins could feel the changing angle of the magnetic force lines, enabling them to establish which part of the Earth's surface they were flying over.

Birds are not alone in possessing a magnetic compass. Butterflies, salamanders, newts, lobsters, bats, whales, turtles, and sharks all have one. In fact, magnetic sensitivity is widespread in the animal kingdom – we are quite unusual in our inability to orientate this way. How the compass works is still not understood, although the active ingredient is believed to be magnetite. In 1979, a team of scientists led by Dr Charles Walcott sensationally discovered particles of this magnetic substance in the head of a homing pigeon, and tiny quantities of magnetite have since been found in many other migratory animals, most recently in the thorax of monarch butterflies.

The advantage of a magnetic compass is that it is unaffected by cloudy weather or changing day length. It can fail during violent electrical storms and sunspot activity, however. There seems to be a link between sunspots and mass beachings of sperm whales, which suggests that their magnetic compass had been disabled.

SHARK SUPERSENSES

Is it true that sharks are mindless killers? It would be more accurate to describe them as swimming computers. Sharks have much larger brains than most other fish relative to their size, with which they process a non-stop stream of information from their environment. Their ability to perceive slight changes in the temperature, salinity, and chemical content of water, "read" wave and current patterns, and detect bands of magnetism in the seafloor, enables sharks to regulate their movements with astonishing precision. Satellite-tracking studies are revealing previously unknown migrations in many shark species: great whites make regular ocean crossings, for example, and basking sharks have been found to travel from British waters to Canada.

Below Unfairly maligned in the media, great white sharks are worthy of respect and admiration for their migratory capabilities. Like other sharks, these ocean nomads have amazing powers of sensory perception.

Mental Maps

The miracle of migration is that it is instinctive. A migratory animal's brain is hard-wired with a preset "program" that tells it to proceed in a specific direction in a certain way at a given moment in time. Naturally there are exceptions, but in general most species are born migrants, inheriting a full set of instructions from their parents that will in time enable them to complete the task.

It used to be assumed that migratory journeys were essentially the product of practice and experience. The problem with this theory was that it could not adequately explain how young animals left by their parents at an early age manage to migrate unassisted. For instance, most species of bird carry out their maiden migration with no parental guidance. The same is true of a host of other animals, many of which never even meet their parents.

We have now grasped that migration is an example of genetically predisposed behaviour. It is controlled by complex inner drives passed down from generation to generation, although this does not preclude individual migrants from getting better due to their life experiences; as might be expected, older migrants are indeed the most accurate navigators. Nevertheless, the key debate today is not about nature versus nurture. Regardless of whether instinct or learning is more important, some of the most exciting new research is focusing on the type of inbuilt map that migrants might use.

BRAIN POWER

True navigation, by definition, ought to require a mental map of some kind, because for it to be possible an animal would have to possess a benchmark against which it is able to establish its current position relative to its ultimate destination. However, scientists have differing views about the extent to which animals are guided by a detailed map outlining the route. Some propose that there is a different navigational system yet to be discovered. For example, animals may resort to a "gradient" system, whereby they repeatedly compare two changing features of the planet's surface against each other.

One thing is certain – if a migrant were to rely on a mental map to reach its goal, it would have to store a vast amount of information in its head. To check its progress during a journey, it would need to refer to this data bank repeatedly using the full range of orientation clues available, both visible and invisible. Is this possible? We simply don't know. However, there is no doubt that the analytical power of a migratory animal's brain is a match for any portable GPS device. Studies have shown that migratory songbirds have a larger hippocampus (the part of the brain that deals with spatial awareness, learning, and memory) than closely related species that do not migrate. It seems that travel really does broaden the mind.

LEARNING THE ROPES

The young of some migratory animals remain with their mother – or both parents – for a considerable time, building up knowledge of a large area and undertaking an entire migration cycle in their company before finally becoming independent. This behaviour is seen in geese, swans, and cranes, which travel in close-knit family units comprising both parents, and it is also found in various whales and hoofed mammals whose offspring migrate with their mother while still suckling. Since the new generation has the opportunity to acquire the established migratory pattern from the previous one, these animals are exceptions to the rule that a well-developed migratory capability is present at birth.

Facing page What can this loggerhead turtle be thinking as it swims towards its target? The truth is that we have no way of knowing if it has a mental map or how it might operate.

GENETICS AND MIGRATION

Eurasian cuckoos provide a classic case study for researchers investigating the genetic basis of migration. These dove-sized birds are brood parasites: that is, the females lay their eggs in the nests of other birds. Yet despite being raised by foster parents belonging to a different species, the young cuckoos duly set off at the end of summer to migrate southwards along much the same routes used by their true parents a month or two earlier, joining the rest of their species on the wintering grounds in East Africa and Southeast Asia. In other words, juvenile cuckoos simply “know” what to do. They must be following a migratory schedule inherited from their parents, complete with a preferred direction of travel.

Left Young Eurasian cuckoos never meet their true parents, yet have no problems finding their winter range. This one is growing up in a sedge warbler nest.

Myths and Mysteries

A REVERED FAMILY

Few migratory species have so inspired people across place and time as cranes. All around the globe, these elegant, heron-like birds are regarded as symbols of hope, fertility, long life, and seasonal renewal. Their wanderlust has much to do with their hold on our collective imagination. Most members of the crane family make long journeys, and they are remarkably faithful birds, turning up at the same locations each year with the same partner. Parties of cranes announce their return with hauntingly beautiful cries and graceful dancing displays. In Rajasthan in northern India, villagers hand out grain to welcome back the wintering flocks of demoiselle cranes, and festivals to celebrate the passage of whooping and sandhill cranes are held in 15 US states. Crane carnivals also take place in China, Japan, and Sweden.

Below Flights of migrating sandhill cranes are a time-honoured cultural icon in North America. Their trumpeting calls symbolize seasonal renewal.

Human beings first became aware of animal migration in the Stone Age, since when it has aroused a mixture of curiosity, wonder, and even religious fervour. Every society developed stories to explain the mysterious workings of this natural phenomenon, while rituals and festivals grew up to mark the endless comings and goings of migratory animals.

The earliest evidence of human familiarity with migration is rock art portraying animals moving across the African savanna. Some of the images, daubed on cave walls and under cliff overhangs throughout continental Europe, date back at least 20,000 years. These artworks were produced by nomadic hunter-gatherers, and could have served as a kind of field guide to potential food or as a visual record of good hunting areas.

Above This fanciful woodcut from Olaus Magnus's medieval treatise, *Historia de Gentibus Septentrionalis et Natura*, published in 1555, shows fishermen hauling in a net laden with fish and hibernating swallows.

Animal migration also played an important role in both the secular and sacred life of the ancient Egyptians. The fertile Nile Valley lies on a major migratory flyway linking Europe and sub-Saharan Africa, and this great civilization knew that the ever-changing cast of water birds visiting their kingdom was somehow connected to the passing of seasons and to the rearrangement of the Sun and stars over the course of the year. The tomb paintings of Ancient Egypt accurately depict more than 70 types of bird, including numerous migrants from further north, such as cranes, sandpipers, ducks, and three species of geese that nest in the Arctic – the bean, white-breasted, and red-breasted.

Many other early peoples became observers of migratory animals, usually because they relied on them for food. Native North American societies depended on their ability to find caribou, bison, waterfowl, or whales at the right time, and these creatures were central to their religious beliefs. Elsewhere, the seasonal movements of birds, game, and even insects formed part of agricultural calendars, as the arrival of certain species was taken as a favourable omen to plant crops.

OUTLANDISH THEORIES

The philosophers of Ancient Greece were the first to develop something approaching a scientific theory of migratory behaviour in animals, although their conclusions often owed much to the realm of alchemy. Aristotle (384–322 BCE) realized that a number of birds were migratory, but his ingenious explanation for the sudden disappearance of summer migrants such as swallows and warblers was that they had magically morphed into different species present in winter. His idea of transmutation persisted well into the Middle Ages in Europe.

According to one theory, widely believed until as recently as the 1800s, when migratory birds vanished they hibernated in mud at the bottom of ponds and lakes. Such ideas may seem far-fetched to us, but were reasonable attempts to make sense of what must have seemed a very strange occurrence. Early naturalists also struggled to account for the fact that adult European eels were plentiful in rivers, when no eggs or young fish could ever be found. The riddle was finally solved in the 1920s, when the species' Sargasso Sea spawning grounds were located.

SOUTH PACIFIC ODYSSEY

Several episodes of human emigration are said to have been prompted by animals on the move. This may sound romantic, but could help to explain the colonization of New Zealand. The first people to reach New Zealand were seafaring tribes from Polynesian islands in the South Pacific, who probably landed there in a flotilla of canoes about 1,000 years ago. Could they have been guided by sightings of migratory seabirds? Each year from September to November, large flocks of shearwaters and petrels stream south from the North Pacific to their breeding grounds around New Zealand and Australia, returning several months later.

SCIENTIFIC PROGRESS

During the 18th and early 19th centuries a new science emerged, concerned with the study of plants, animals, and other living things, and how these organisms interact with each other and with their environment; it would come to be known as ecology. Carl Linnaeus (1707–1778), one of the fathers of modern ecology, devised a rigorous system to classify species, paving the way for systematic research into animal behaviour in the wild. Naturalists organized expeditions all over the world, discovering and naming thousands of species and slowly adding to knowledge of migratory patterns. However, the greatest breakthroughs in migration studies would occur in the 20th century.

Above Rock paintings, such as this one from Damaraland in Namibia, Africa, are the oldest known representations of animal migration and provide clues about the historical ranges of migratory species.

Above Some blackcaps no longer migrate from their European breeding grounds to Africa. Studies have shown that the non-migratory blackcaps have smaller wings than their migratory relatives – an example of evolution in action.

Origins

Migration is a dynamic, constantly evolving process. Although individual migrants adhere to a predictable route and routine, their species' migrational patterns will, over time, alter in response to changing threats and opportunities in the world around them. Migration is but one example of the relentless adaptation of species to their environment as they compete for living space and resources.

The Earth has been in a state of flux throughout geological history, with profound consequences for animal and plant life. Over millions of years, the realignment of continents and the appearance of new coastlines, land bridges, and mountain chains have made animals continually modify their migration routes. These forces are felt by marine species, too. For example, the marathon migrations undertaken by humpback and grey whales were probably far shorter in the past, since when continental drift has pushed their cold-water feeding and warm-water breeding areas further and further apart.

Arguably the most important factor influencing migration has been the cycle of ice ages, which has driven the habitat that each species needs back and forth across the surface of the planet over time. Many of the long-haul animal migrations seen today have their origins in the global warming that brought the last ice age to an end, approximately 10,000 years ago. As the Northern Hemisphere's ice sheet started to recede towards the Arctic as temperatures rose, so too did the tundra zone that formed the sheet's southern boundary. Species that visited the tundra to breed, including caribou, sandpipers, swans, and geese, found that the journey to and from their winter quarters, in milder climes to the south, became slightly longer with each generation.

EVOLUTION OF MIGRATION

Migratory behaviour is not evenly spread throughout the biological world – it tends to evolve in certain biomes and animal groups but not in others. We have already seen that true migration is rare among the primates (*see* pages 14–15), despite there being about 230 living

Above Sperm whales belong to the toothed whale group and prey on squid. Over time, the species has evolved a system of partial migration: only the large bulls travel long distances, while mothers and calves remain in the tropics all year.

WHALE PUZZLE

One of the great migration riddles is why most baleen whales swim to the tropics to breed, which are food deserts as far as these planktivorous leviathans are concerned. Surely it would make more sense to remain in nutrient-rich polar waters all year? Indeed, this is precisely what the bowhead whale does, spending its entire life in productive Arctic seas. It could be that thousands of years ago the other species of baleen whale began traveling to lower latitudes because they enjoyed better breeding success. Pregnant females require calm, sheltered coasts to give birth, and these are easier to find in the tropics. In addition, there are fewer prowling orcas, and shallow tropical waters are warm and highly saline, so serve as perfect nursery grounds for newborn whale calves lacking the insulation and buoyancy provided by thick blubber.

species. The same can be said of parrots: of the 350 species known, few carry out long seasonal movements, and only two ever cross the sea on a regular basis. This is hardly surprising, because both primates and parrots have a strong preference for tropical forests, where the constant high temperatures and abundant rainfall ensure luxuriant plant growth all year and there is little incentive to travel.

By contrast, tropical grasslands experience alternating wet and dry seasons, with dramatic climatic extremes. As a result, vegetation growth occurs in spurts, between which the rain-starved landscape turns golden brown and food is hard to come by. The dominant groups found in the savanna biome are herbivores, particularly even-toed ungulates (Artiodactyla), and it is no coincidence that most members of this great mammal order lead highly migratory or nomadic lifestyles.

GRADUAL CHANGE

The genetic basis of migration means that, like any inherited trait, it develops by the process of natural selection. Each species' inbuilt migratory "program" adapts gradually, over successive generations, to improve the chances of survival. The change is usually almost imperceptible to us. Sometimes, however, it happens fast enough to be measurable within a human lifetime. One notable case is the migration of the blackcap, a small, insect-eating songbird that breeds in woodland in Europe and western Asia, wintering in Africa south of the Sahara. Since the 1970s, increasing numbers of blackcaps have stayed behind to spend the winter in northern Europe, no doubt in response to warmer weather. This non-migratory behaviour probably started by accident, when a few individuals failed to migrate in the normal way, but it is now entrenched in a sub-population of northern blackcaps because it saves them the effort of migration, thus enhancing their survival prospects. Eventually, the populations could evolve into separate species.

Studying Migration

Our understanding of animal migration has come a long way. Now the journeys of migrants as small and fragile as dragonflies can be tracked in great detail using sophisticated transmitters and data loggers, while new analytical techniques let scientists delve into the chemical make-up of migrants themselves.

Until the early 20th century, most observations of migratory patterns in animals came to light through activities such as whaling and big game hunting, or by gathering specimens for display in natural history collections. But in the 1940s and 50s, as field identification skills and the quality of optical equipment improved, naturalists began systematic studies of animals actively migrating. A network of observatories was established at islands, headlands, and other migration hotspots to log the passage of birds. Today, we live in the age of the online census: sightings uploaded by dedicated armies of observers help to build, in real time, a detailed picture of animal migration in its myriad forms.

To keep up with active migration, researchers take to the skies in everything from microlights to hot air balloons and light aircraft. Aerial surveys are useful for counting elusive, wide-ranging species, particularly grassland herbivores and large ocean-going voyagers such as blue whales, whale sharks, and leatherback turtles. In order to survey a representative area, spotter planes zigzag along fixed transects at a height of about 150 m (500 feet).

IDENTIFYING INDIVIDUALS

One of the simplest methods for studying migration is to mark individual migrants so that they can be recognized. It is then possible to get an idea of their movements by drawing a straight line between points where they are seen. This approach dates from 1899, when Danish teacher Hans Christian Mortensen invented a technique called ringing. Mortensen visited European starling nests and gave each nestling an aluminium leg ring engraved with a unique serial number and return address. If anyone came across one of his birds, they could send back details of its new location and the date it was found.

CHEMICAL ANALYSIS

The newest branch of migration studies involves comparing chemical samples. In isotope tracing, for example, biologists measure the amount of deuterium (a stable isotope of hydrogen). The deuterium level in the plumage of a migratory songbird matches that of the vegetation of its breeding ground, thus serving as an indicator of its origins. The same technique has been used to establish the hatching place of individual monarch butterflies wintering in Mexico.

Since 1899, more than 200 million birds are estimated to have been ringed around the world, of which only a fraction have ever been "recovered" (a euphemism for being trapped, shot, or picked up dead). However, a recovery rate as low as 1 in 300 – the average for small birds – still gives a valuable insight into where migrants go. Alternatives to ringing include labelling with dye and attaching plastic tags to the neck or back, procedures that are equally effective with mammals. There are special tags designed for the flippers of sea turtles. Even insects have been marked successfully, for example by painting the wings of monarch butterflies or sticking tiny labels onto them.

RADAR AND SATELLITE TRACKING

The rapid development of radar after the Second World War enabled actual migratory journeys to be plotted for the first time. Although radar operators were initially puzzled by waves of interference on their screens, referring to them as "angels", they soon realized that these patterns were caused by streams of migrating birds. Modern radar is powerful enough to pinpoint the height, speed, and wing beats of individual birds and bats, while its aquatic equivalent – sonar – can detect shoals of fish moving underwater.

Above A pop-off archival tag for large fish, which will relay data on location, depth, and ambient light and temperature to a satellite system when it has been programmed to do so.

In recent decades, research into animal migration has been revolutionized by satellite tracking, also known as satellite telemetry. The migrants being studied are fitted with tracking devices called platform transmitter terminals (PTTs), which beam signals to orbiting satellites according to a preset cycle, and these in turn relay the information back to computers on the ground. Since the 1990s, transmitters have become progressively smaller and lighter, with a longer battery life and stronger signal (hence range), allowing the progress of migrants to be followed for months. The latest transmitters come in many types, including collars, darts, anklets, and miniature units glued to the thorax of insects, and give updates about an animal's body processes and environment as well as its movements.

In addition to tracking devices, there are long-term data-storage instruments, generally used with fish. Often these are surgically implanted into the fish, but the drawback with internal tags is that to retrieve them the subjects need to be caught again and then dissected. An alternative is to use pop-up archival tags (PATs), which are programmed to detach at a preset time and float to the surface to upload their stored data via satellite, meaning that it is not necessary to recover the gadget itself.

TRACKING SOOTY SHEARWATERS

In 2005, researchers deployed satellite-tracking tags to investigate the dispersal of sooty shearwaters, one of the world's most abundant species of seabird. Nineteen shearwaters were fitted with the tags at their New Zealand breeding colonies. They were then followed through the Pacific as they embarked on a trans-equatorial circuit that took them as far north as Japan, Russia's Kamchatka Peninsula, and Alaska. Each bird's tag transmitted positional data and also registered the depth of the bird's dives and the temperature of its surroundings.

Map A below illustrates the different tracks taken by the shearwaters over a period of 222–313 days: pale blue indicates foraging trips made during the breeding season; yellow indicates paths taken at the beginning of the migration as the birds made the passage north; and orange indicates movements on the wintering grounds and the subsequent routes south. **Maps B–D** represent the complete migrations of three mated pairs in the sample, revealing that the birds trace a figure-of-eight path and stick to a narrow corridor in the central Pacific, probably to take advantage of global wind circulation patterns.

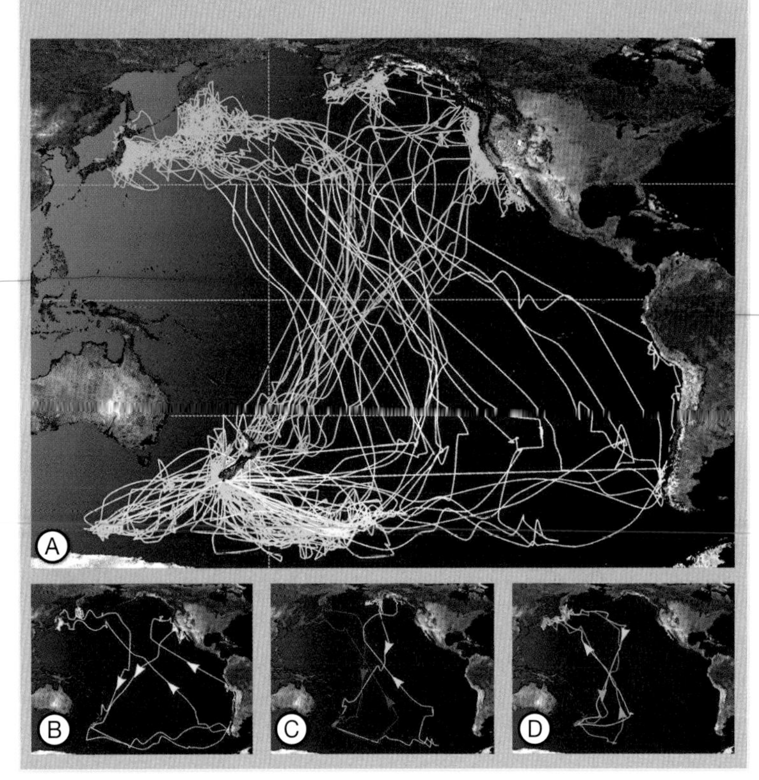

Above left Birds are usually ringed when nestlings for convenience, but on migration they can be caught in a variety of harmless ways, including with mist nets. As its name suggests, a mist net is made of mesh so fine as to be almost invisible.

Left The young caribou in this photograph has been fitted with a geolocating collar that will send out a radio signal for researchers to track the animal. The large numbers can be read through binoculars at long range to keep disturbance to a minimum.

Growing Dangers

Nature sees to it that migratory species and cycles persist despite the dangers they face. But humans have upset the status quo almost overnight in biological terms by dramatically increasing the risk. Animals are confronted with extreme hunting and fishing pressure, a wealth of human-made obstacles, and rampant habitat destruction.

Hunting is one of the oldest forms of human interaction with wildlife, but what has changed during the past 200–250 years is the intensity of the threat and the number of species at risk. Many of the megafauna alive today, including elephants, rhinos, bison, and whales, very nearly didn't make it into the latter half of the 20th century. Buccaneering whaling captains competed to see who could catch the most, while big-game hunters believed that their quarry existed in "infinite numbers" and amassed huge bags on their trips; a single page of a South African game control officer's logbook for 1908 recorded the shooting of 996 rhinos.

Smaller species of game were not spared either. The mass migration of springbok across the plains of southern Africa used to dwarf that of East Africa's wildebeest, until white settlers shot so many that the Trekbokken (as the annual spectacle was called) fizzled out. Now it exists only as a distant folk memory. Herds of wild springbok that were formerly millions strong rarely number more than a few hundred today.

We should know better in the 21st century, yet still the carnage continues. The roll call of migratory species currently threatened by excessive harvesting includes most large tuna, sturgeon, and sharks; Atlantic cod and several of its relatives in the order Gadiformes; six of the world's seven species of marine turtles; numerous gamebirds and waterfowl; and Asian antelopes such as saiga and chiru. Sometimes the victims are caught unintentionally – for example, it is estimated that 100,000 albatrosses are killed each year on longline fishing hooks set for tuna. That's one every five minutes.

Left Much of the world's 1.5 billion hectares (3.7 billion acres) of arable land is monoculture – a sterile, featureless landscape of little or no use to migratory animals.

Left Power lines kill untold numbers of migratory birds, especially large, slow-flying species such as raptors and storks. Utility companies are increasingly held liable for this situation, and the deadly cables are being routed away from important flyways.

ARTIFICIAL OBSTACLES

Wilderness is at a premium in our increasingly crowded world, leaving less space for animal migration to continue unimpeded. On land, migrants as large as elephants and as small as toads are finding their ancient routes blocked by fences, highways, and urban sprawl. Nowhere is sacrosanct – even remote tracts of the High Arctic are becoming cluttered by oil and natural gas pipelines, which drive a wedge through the traditional pathways followed by caribou. In the skies, migrating birds have to contend with lethal power lines, and dodge the gleaming forests of wind turbines that have sprouted on hilltops and in coastal waters.

There is growing evidence that telephone masts and radio and TV transmission towers pose an additional danger: their flashing lights, and the magnetic radiation they produce, seem to impair birds' sense of direction. Studies on bobolinks, a small species of New World blackbird, have shown that these nocturnal migrants are easily confused by bright red lights of the type found on transmission towers, heading straight at them. The size of this problem could be immense – one survey suggested that the towers may be responsible for 3–4 million bird deaths per year in the USA alone.

BROKEN LANDSCAPES

One of the main reasons migratory animals feature prominently in lists of endangered species is their need for interconnecting corridors of suitable habitat to enable them to move from one piece of their range to another. At a local level, migratory pathways can be erased by something as simple as the removal of a wood, meadow, pond, or ditch. At a regional level, there is wholesale conversion of natural, seasonal habitats to new uses. Thirty-five per cent of the Earth's land surface has already been usurped by agriculture. About a further one-third is classed as urban. Rivers have been dammed, diverted, and drained. This is one of the greatest ecological changes in the planet's history, and it has left migratory animals, particularly mammals, with vast no-go areas.

Part of the problem is that many national parks and reserves were planned and created long before we knew the location of important migratory corridors. A random patchwork of small, isolated refuges is not enough – what is required is an interconnected network of protected areas in both public and private ownership. Only a global approach to conservation can protect migratory species that cross international and regional boundaries.

LOST FOR GOOD

The fate of the passenger pigeon, shot and poisoned to extinction in the wild by 1901, provides a tragic demonstration of how even abundant species can be wiped out. These pigeons nested in dense colonies in the forests of the eastern USA and southern Canada, and after breeding migrated to southern states and Mexico for the winter. On their journey through the Midwest, they formed probably the largest gatherings of birds ever seen – some flocks were reputed to have been up to 1.6 km (1 mile) wide and 500 km (300 miles) long. But the migrating pigeons were slaughtered on such a scale – for meat, for use as fertilizer, to protect crops, or simply for sport – that in the end the unthinkable happened. In 1914, the last passenger pigeon on Earth died in Cincinnati Zoo. Its species' demise is a reminder that species are always vulnerable to hunting if they congregate in a few places.

Above Once the most numerous bird in North America, the passenger pigeon was annihilated within several decades, warning us that the survival of migratory species should never be taken for granted.

Above China's Three Gorges Dam, which spans the Yangtze River in Hubei province and is the largest of its kind, has become a symbol of the environmental devastation caused by massive hydroelectric schemes.

Below Olive ridley turtles are under greater pressure today than at any time in their long history. Man-made global warming could be the final straw if, as predicted, it results in rising sea levels, increased sand temperatures on beaches, and changes to ocean currents, all of which would be disastrous for the turtles.

The Beginning of the End?

Climate change is set to become the most severe threat to migratory species. Rising temperatures and increasingly unpredictable weather have begun to damage the seasonal habitats used by migrants and are disrupting long-established migration patterns. It remains to be seen if we have the willpower to reverse this potentially catastrophic process.

By relentlessly burning fossil fuels, we have unleashed what could develop into, or may already be, a mass extinction event – the sixth in the history of the Earth. Thousands of species are likely to slide towards oblivion unless urgent remedial action is taken. There is a growing scientific consensus that we have until about 2015 to prevent greenhouse gas emissions passing the critical tipping point at which global warming becomes unstoppable.

OUT OF TIME

Migratory species are more vulnerable than most to a volatile climate, because they need to time their journeys to take advantage of short-lived seasonal resources, such as supplies of food and water. They often pass through different habitats over the course of a year or lifetime, and have to be in each one at exactly the right moment. Being "off the pace" can spell disaster. In addition, climate change means that migratory journeys are inherently more dangerous; it can cause deserts to spread, ocean currents to shift, sea ice to break up, and storms to become more violent. Due to this sensitivity to change, migrants act as an early-warning system for the health of our planet, and the prospects do not look good.

THE EVIDENCE SO FAR

Worrying trends from a variety of biological indicators show that all is not well with some migratory animals. For instance, a census of grey whales in the East Pacific in 2006–2007 found that the changing climate was affecting their breeding success and general health. Unusually high numbers of whales were travelling in poor body condition, suggesting that they could not find enough invertebrate prey in their key Bering Sea feeding grounds; the sea may have become too warm, reducing productivity. At the opposite end of the world, the Southern Ocean is also heating up, with equally serious consequences for marine life adapted to cool conditions, from tiny krill to species at the top of the food-web such as albatrosses, seals, and whales.

DISAPPEARING TUNDRA

If global warming continues unchecked, Arctic tundra will be among the first habitats on land to suffer. A sustained shift to milder winters and longer summers is predicted to trigger a widespread thaw of permafrost (the layer of permanently frozen soil), enabling the coniferous forests of the boreal zone to advance north and eat into this open, treeless wilderness. In the worst-case scenario, migrants that visit the tundra to raise young, including waders, waterfowl, and massed herds of caribou, would be left with nowhere to go.

Above It is anticipated that conifers will invade three-quarters of the red-breasted goose nesting area by 2070.

Fewer monarch butterflies are surviving their period of winter torpor in the Mexican mountains, probably due to more frequent storms and a shift to warmer, drier winters, triggered by the La Niña weather pattern. Usually the species is able to bounce back quickly the following year, but factors such as pesticide use and habitat destruction in the butterflies' American summer range have dramatically increased the pressure on their populations. It is the combination of climate change with other forms of environmental degradation that is causing problems for the monarchs. The same is true for many other migratory animals. Sea turtles are ancient survivors that have lived through over 150 million years of climate change, but when coupled with hunting, marine pollution, and habitat destruction, it may be more than they can bear.

Migrants can and do respond to climate change. For example, some migratory birds arrive on their breeding grounds earlier, while others leave later. But in many cases they are not reacting quickly enough. Globally, spring events have advanced by 2.3 days per decade, with the result that in temperate latitudes, habitats are moving towards the poles; migrants may be left stranded, in the wrong place at the wrong time.

Above This herd of plains zebra is moving at full gallop. Zebra families travel hundreds of miles a year as they move between islands of rain-ripened grass.

Migration over Land

The world's land animals have undertaken epic migrations since time immemorial. They may trek for weeks before finally arriving at a safe haven, across such unfriendly terrain as boggy tundra, ice sheets, hot deserts, rocky mountains, or even active volcanoes. Unlike aquatic creatures, which hitch a lift on fast-flowing currents, or airborne travellers that ride the wind, terrestrial migrants must journey on foot. As a result, their excursions tend to be shorter than those by water or air, but the sheer number of animals on the move together is often astounding.

Caribou

Caribou travel further in a year than any other land mammal. Every summer they faithfully return to the same traditional calving grounds on the treeless tundra, then trek south to sheltered forests for the winter. The great tide of caribou crosses mountains, rivers, and lakes, and may take days to pass a single point.

Caribou are the champion migrants of the deer family. These perpetual nomads roam freely over some of the largest wilderness areas on Earth, throughout the transitional zone where coniferous forests of the boreal zone merge into the bleak, windswept tundra of the High Arctic. Their summer and winter ranges are usually 160–800 km (100–500 miles) apart, but when local movements are taken into account, the total distance covered is far greater. A radio-tracking project in Quebec, Canada established that some caribou travel as much as 6,000 km (3,700 miles) a year.

European and Russian caribou, known as reindeer, were once widely domesticated. They are still herded today by several nomadic peoples, such as the Sámi of Lapland and the Nenets of Siberia, who travel with the reindeer herds, sleeping in reindeer-skin tents and carrying their belongings on sleds hauled by the animals.

LOCAL VARIATIONS

Wild caribou herds vary considerably in size, from a few thousand to a hundred thousand individuals, or even more; there are three herds estimated to be 500,000 strong. The distance travelled by different herds varies too, depending on the topography and weather. As a rule, northerly herds tend to be largest and migrate furthest. The real long-haul migrants, often referred to as "barren-ground" caribou, are those that visit the tundra in summer.

One of the best-known of all barren-ground herds is named after Alaska's Porcupine River, along which the caribou migrate in spring and autumn. The Porcupine herd has been studied for many years, using a variety of survey methods including satellite telemetry. In 2007–2008, the herd contained about 125,000 animals, of which nearly 50 were wearing satellite-transmitter collars. The collars provide regular location fixes accurate to within 1,000 m (3,500 feet), enabling the herd's movements to be plotted in real time.

YEARLY CYCLE

The Porcupine herd winters in parts of Yukon Territory and in Alaska south of the Brooks Range, and migrates north to traditional calving grounds on the coastal plain and foothills of the Arctic National Wildlife Refuge. But the herd's annual migration is more than a one-way journey between two areas: the herd's year can be divided into eight seasons, based on changes in snow cover and type of forage available.

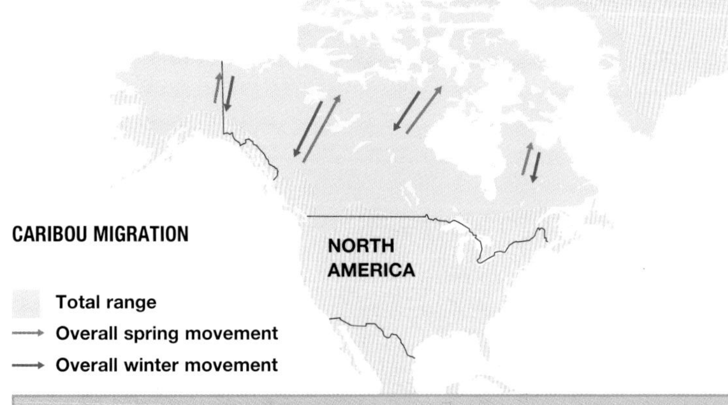

Below During the harsh winter months, caribou eke out their existence by foraging beneath the snow for lichen.

CARIBOU MIGRATION

- Total range
- Overall spring movement
- Overall winter movement

MIGRATION PROFILE

Scientific name	*Rangifer tarandus*
Migration	From tundra in summer to boreal forests in winter
Journey length	Up to 800 km (500 miles) each way
Where to watch	Arctic National Wildlife Refuge, Alaska, USA
When to go	June–July

BIRTHING MIGRATION OF THE PORCUPINE CARIBOU HERD

- Core calving areas
- Extent of calving
- Spring migration route
- Herd range

CANADA

ALASKA

In winter, the staple food of caribou is a type of lichen called reindeer "moss", revealed by using antlers and hooves to dig through the snow. In March or April, by now thin and restless, the caribou head north in search of better grazing. They walk in single file, treading in each other's tracks to avoid trudging through deep snow, pausing to browse the scant supplies of lichen and sedges as they go. With this highly economic gait, they can cover up to 50 km (30 miles) a day, yet expend minimum energy. The caribou are thought to navigate using visual landmarks and a magnetic and sun compass – one study found that they keep to a narrow migratory corridor no more than 15° wide.

By late May, they have reached lush meadows carpeted in new-growth cotton grass, and at last can feed in earnest. The herd's heavily pregnant cows arrive first, having pushed ahead of the bulls and juveniles. Years of data show that almost all births occur between June 1 and 10. The sudden abundance of baby caribou "swamps" predators – mainly wolves and brown bears – minimizing the number they are able to take; it also enables the whole herd to move on together. Young caribou can already outrun an Olympic sprinter when only a day old.

By early July, the warm, damp tundra air is thick with clouds of mosquitoes and warble flies, forcing the harassed caribou to band together and seek refuge by the coast or on ice fields, where cool breezes keep the insects at bay. Despite these tactics, an adult caribou will lose perhaps about 1 litre (2 pints) of blood to biting insects for every week it spends on the tundra. The caribou leave the coast towards the end of July and wander southwards to upland areas, where they remain until September or early October. Finally, they press on to their rutting grounds, and from there to their forested winter ranges to begin the cycle again.

WOLF HUNTS

The chief predators of caribou are Arctic wolves. These thick-furred, almost white wolves – the world's largest subspecies of grey wolf – are responsible for up to 70 per cent of calf deaths and also kill many of the old and sick caribou. Wolf packs can mount surprise attacks on the caribous in their forested winter territory, but in other seasons they track migrating herds across the open tundra, where there is no chance of ambush. The final, flat-out chase can go on for 10 km (6 miles). Wolf cubs are too weak to take part until about 10 months old, the pack members will feed them later on regurgitated meat if the hunt is a success. Unattended cubs may be killed by bears or by wolves from other packs, so summertime caribou hunts can end in disaster for the wolves.

Right Arctic wolves doggedly follow caribou herds across open tundra for several days or weeks, waiting for the right moment to strike.

Facing page This aerial view shows a herd of Alaskan caribou tramping across the soft ground of the tundra towards the feeding grounds further north, where they will fatten up before the coming winter.

Polar Bear

Winter is a time of plenty for polar bears, when they roam the frozen Arctic Ocean in search of seals. But as their icy hunting platform disintegrates in summer, the more southerly bears are forced ashore. Often they must swim many miles to find land.

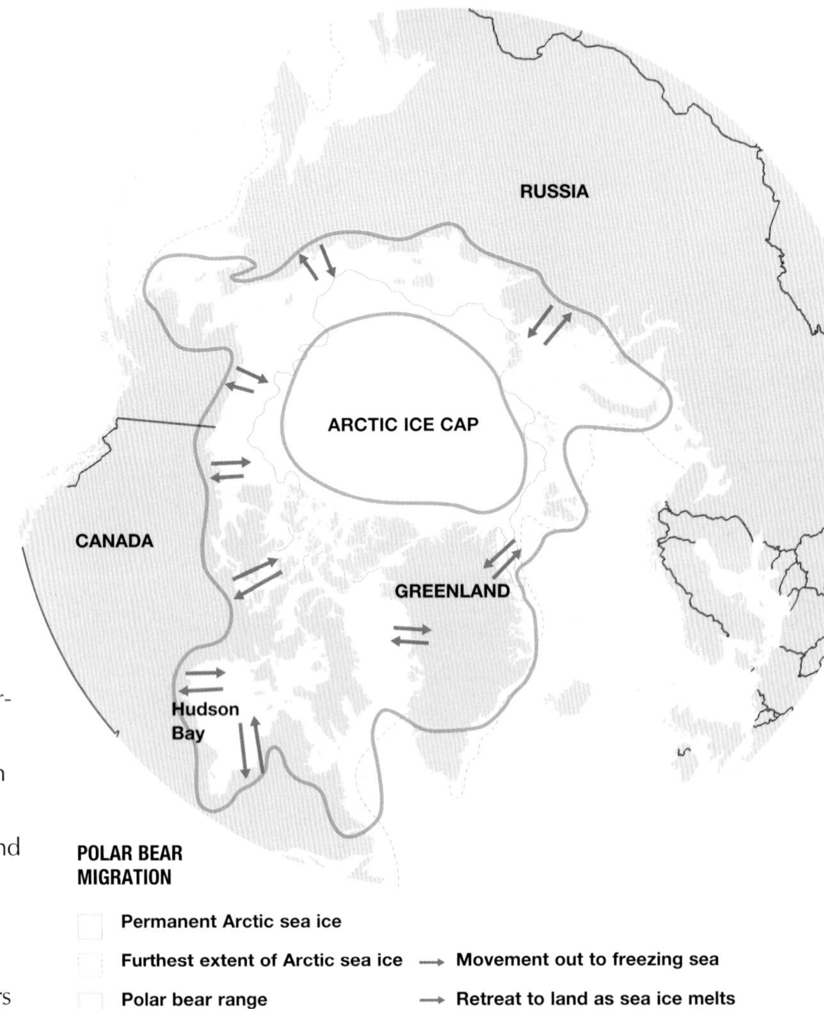

POLAR BEAR MIGRATION

- Permanent Arctic sea ice
- Furthest extent of Arctic sea ice
- Polar bear range
- → Movement out to freezing sea
- → Retreat to land as sea ice melts

Polar bears are the world's largest terrestrial carnivores: 10-year-old adult males weigh up to 800 kg (1,750 pounds), roughly the same as a small hatchback. They top the Arctic food chain and, like many "apex" predators, travel hundreds of miles a year to find prey. Sea ice is a mobile habitat that drifts with ocean currents and freezes differently from year to year, making the bears' home ranges huge. Even a small range covers 50,000 sq km (20,000 square miles), while the biggest are about seven times this size – equivalent to the entire state of New Mexico, USA. Unlike many other carnivores, bears are not territorial – their ranges overlap and have vague limits.

THE FROZEN FRONTIER

Polar bears live between the permanently frozen polar ice cap and the tundra region – contrary to popular belief, there are no bears in the icy wastes near the North Pole. Their stronghold is a tremendously rich ecozone known as the Arctic's "ring of life". This is a maze of polynyas (areas of open water hemmed in by ice but which stay clear all year) and leads (water channels and cracks in the ice) that runs parallel to shorelines throughout the Arctic. It offers superb hunting for the bears.

MIGRATION PROFILE	
Scientific name	*Ursus maritimus*
Migration	Southern populations come ashore in summer and return to sea ice in winter
Journey length	Up to 1,125 km (700 miles) each year
Where to watch	Churchill, Manitoba, Canada
When to go	Mid-October–November

Left Driven by hunger, bears scavenge at a rubbish dump on the outskirts of Churchill, Manitoba in Canada. Due to shortages of their usual prey, the bears have become increasingly brazen in recent years, sometimes venturing into the town itself.

BEARS AND PEOPLE

The town of Churchill, nestled on the western shore of Hudson Bay, Canada, bills itself "Polar Bear Capital of the World". Every autumn, about a thousand bears gather nearby to wait for the Bay to freeze over – the largest, most southerly gathering of this species anywhere. The bears outnumber the townfolk and since the 1980s have provided most of them with a living. People flock to Churchill to watch the bears, which are now so habituated to this annual ritual that it is possible for snow buggies to approach within feet. However, this model example of ecotourism is threatened by global warming. In summer the Bay's pack ice breaks up three weeks earlier than it did 30 years ago, and the shorter sea-ice season gives the bears less time to hunt seals. As the ice disappears, the bears are becoming fewer, thinner, and more aggressive.

Ringed seals are their main target, particularly the pups, whose obese bodies are as much as three-quarters fat. Other marine prey includes bearded seals, narwhals, and young beluga whales and walruses, but these are considerably larger and more difficult to catch. All winter, bears hunt seals by staking out their breathing holes in the ice, waiting patiently until one surfaces for air and then yanking it out of the water. The seals are especially vulnerable during their breeding season, from mid-March to April, when newborn pups shelter in lairs hollowed under ridges of ice and wind-blown snow. The bears' extraordinary sense of smell can guide them to a hidden pup from 5 km (3 miles) away, but even so, fewer than one in three of their smash-and-grab raids is successful.

The bears devote most of their waking hours to hunting, bingeing on seal blubber to lay down fat reserves for the leaner summer months ahead. By June, the seal pups have left their nursery dens for open water, and the rising temperatures create ever-increasing gaps in the ice that the bears must swim across. These powerful swimmers are able to paddle for hours if necessary and have been spotted 95 km (60 miles) from the nearest shore. Eventually, however, the fragmenting ice floes are too small and scattered for efficient hunting, driving the bears to dry land.

Summer paradoxically is a difficult time for polar bears. While on land, the bears become uncomfortably hot in their dense fur coats and are reduced to eating berries and scratching around for roots. They catch a few seabirds and lemmings too, but these are meagre pickings for such formidable hunters, and the bears grow hungry. By the end of October, the near-starving bears are desperate for the sea to refreeze so they can trek after seals once more.

Facing page Polar bears live in an extremely dynamic environment. Because sea ice is constantly on the move, so are the bears.

SOUTHERN MIGRANTS

The most migratory polar bear populations are in the south, where the ice breakup is more widespread and longer lasting. They include the bears of Hudson Bay and Labrador in Canada, the southern half of Greenland, the Bering Sea, and southern portions of the Chukchi and Beaufort seas. Broadly speaking, these bears migrate along a north–south track, as the pack ice retreats northwards in spring and advances southwards in autumn, although there are many local variations.

The Beaufort Sea population is very well documented, due to ongoing studies in which female bears are fitted with satellite telemetry collars. These data have revealed that the bears do not wander at random; rather, they follow the shoreline and channels through the ice. They are fairly sluggish migrants, seldom moving more than 50 km (30 miles) a day.

A NEW COLD WAR

Today, polar bears find themselves on the front line of environmental politics. Global warming has already shortened their feeding season by shrinking the Arctic's sea ice, with the result that, on average, bears weigh 80–90 kg (175–200 pounds) less than 15 years ago. A 2007 US Geological Survey report stated that unless there were drastic cuts in greenhouse gas emissions, two-thirds of the global population of polar bears – including all those in Alaska – would disappear by 2050.

Above A mother guides her two young cubs across the frozen sea in search of seals. Her offspring's chances of survival will depend on her ability to navigate the labyrinth of floating ice and open water channels to locate the most productive hunting grounds.

Life at the Top

Himalayan peaks are a formidable barrier to life, with thin air, hostile terrain, and sparse vegetation, yet migration enables a surprising number of animals to eke out an existence in these lofty cold deserts.

Every continent apart from Australia and Antarctica has native herbivores that migrate up and down mountainsides with the change in season. Herders, too, guide their flocks from alpine summer pastures to lower valleys in winter – an ancient type of livestock husbandry called transhumance. By shifting a few thousand feet downslope, animals and people alike can obtain as much climatic benefit as if they had travelled hundreds of miles towards the equator.

The Himalayas are an evolution hotspot, comprising many isolated ranges with different climates created by the uneven influence of the monsoon rains, and therefore support a great diversity of high-altitude mammals. The majority undertake vertical seasonal migrations in response to snow coverage. Among them are sheep, including the bharal and argali; several deer; the goat-like markhor and Himalayan tahr; and the bizarre-looking takin, which resembles a cross between a musk-ox, wildebeest, and buffalo. All possess long, shaggy coats as protection against the bitter cold.

The main predator of these stocky, sure-footed herbivores is the endangered snow leopard, which today numbers only about 5,000 individuals, scattered thinly across the remote highlands of 12 countries in Central Asia. Seldom seen, the elusive cats roam widely to track their prey, climbing to 5,000 m (16,500 feet) or more in summer and descending to 3,000 m (10,000 feet) in winter, to the edge of juniper forests and scrub at lower elevations.

Left Many Himalayan herbivores, such as these tahr photographed in Nepal, migrate uphill in summer to graze the scant vegetation at higher elevations.

Below left In winter takin retreat to forested ravines and lee slopes sheltered from the brunt of icy winds. They follow well-used trails that hug the contours of the land.

Below right Snow leopards mirror the seasonal altitudinal movements of wild sheep and goats, their favourite prey. Food is scarce, but on average they kill one large animal every two weeks.

Dall Sheep

Dall sheep inhabit some of the most hostile, rugged terrain in North America. They shift habitats as the seasons change, visiting high alpine pastures in summer and spending the winter on snow-free slopes lower down.

Renowned for their ability to scale near-vertical crags and for the violent battles between rutting rams, mountain sheep embody wilderness in the popular imagination and have an intriguing history in North America. During the last ice age, which began about 110,000 years ago, vast glaciers pushed south from the polar icefields, stranding the sheep in two main ice-free regions. Those in Alaska evolved into a species with slim, widely curled, pointed horns, known today as thinhorn sheep; those in the Rockies and southwestern deserts developed thicker, blunt-tipped horns, hence their name bighorn sheep.

Dall sheep are a subspecies of thinhorn sheep with pure white coats, and honour the American naturalist and explorer William Dall (1845–1927). Their homeland is the continent's remote northwestern backbone, encompassing mountain ranges in Alaska and across Yukon, Northwest Territories, and British Columbia. Full-grown rams, those at least four years old, are a magnificent sight. They stand 81–107 cm (32–42 inches) at the shoulder, and in their prime during autumn weigh 82 kg (180 pounds) on average – approximately half as much again as the lighter-built ewes.

ALTITUDINAL MIGRANTS

Many of the world's wild sheep, including Rocky Mountain bighorns and mountain sheep from Europe and Asia, as well as Dall sheep, undertake altitudinal migrations. That is, their seasonal movement is more a journey between heights than between places. Such an itinerant existence brings obvious benefits. Upland meadows have fewer predators than lowland plains, enabling these agile animals to raise their young in comparative safety, and there are few other ungulates, or hoofed mammals, to compete for grazing. However,

Facing page With impressive bravura, Dall sheep can skip across precipitous walls of rock, enabling them to escape their enemies and reach the highest mountain pastures in summer.

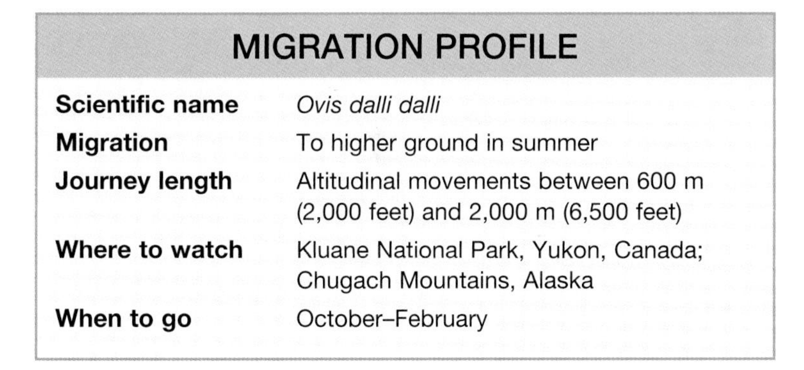

MIGRATION PROFILE

Scientific name	*Ovis dalli dalli*
Migration	To higher ground in summer
Journey length	Altitudinal movements between 600 m (2,000 feet) and 2,000 m (6,500 feet)
Where to watch	Kluane National Park, Yukon, Canada; Chugach Mountains, Alaska
When to go	October–February

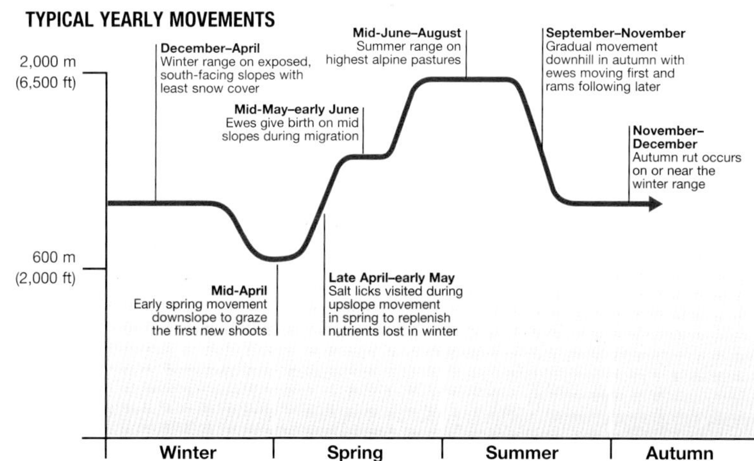

naturally there is a price to pay: the need to climb sheer rock faces imposes a strict size limit on sheep; and the grass at higher elevations is of poorer quality.

Dall sheep stay above the timberline all year, in segregated flocks called bands. The mature rams form bands up to 15 strong, and the ewes gather in larger bands accompanied by juvenile rams and (in early summer) lambs. Bands mingle for the November–December rut, then separate again. Every band has its own distinct pattern of movements, dependent on the severity of the local climate and the lie of the land, but everywhere there is a general trend to the migrations.

Sometime in late April or May, the sheep set off from their winter quarters to follow the spring melt uphill, moving onto grassy mountainsides and sedge meadows newly exposed by the retreating snow. The ewes, which have been pregnant all winter, give birth in the second half of May or early June, choosing a secluded spot in a ravine or under a rocky outcrop. Each produces a single lamb every other year, and within days the mothers and their offspring will rejoin their group.

As summer progresses, the sheep climb to alpine pastures, often as high as 2,000 m (6,500 feet) in Alaska's Brooks Range. By early autumn, as supplies of grass are exhausted, the sheep must descend to subalpine slopes and valleys, and hunger forces them to supplement their diet with lichens, dwarf willows, and mosses. Finally, as the snows arrive, they seek refuge on patches of south-facing grassland, particularly exposed ridges where the wind blows away snowdrifts to reveal the turf beneath.

INHERITED RANGES

Dall sheep are not great wanderers. Usually they remain in the same, rather small, home range for their entire life. An ideal range includes a mix of open grazing areas, cliffs (to escape predators), sheltered gullies, south-facing slopes, and mineral licks (especially important in spring, for replenishing nutrients lost during the winter). This patchwork of habitats is connected by a well-worn network of sheep tracks, used year after year. Field studies suggest that young rams tend to follow older, heavy-horned rams, and young ewes likewise follow experienced, lamb-leading ewes; in this way, crucial home-range knowledge and migration routes pass to the next generation.

Unlike species such as caribou (*see* pages 44–45) and moose, Dall sheep seem extremely reluctant to disperse and pioneer new areas – hunters have long used this predictability to their advantage. Unfamiliar canyons and wooded valleys are dangerous places, where wolves and coyotes prowl, so there is good reason for the sheep to be cautious.

GLACIAL SPLIT

A second subspecies of thinhorn sheep is found to the south of Dall sheep, in parts of southern Yukon and northern British Columbia. These are known as Stone sheep (*Ovis dalli stonei*) and are almost completely black but otherwise identical to their northern counterparts. How did this contrasting coloration arise? One hypothesis proposes that about 10,000–20,000 years ago, glaciation carved out two separate populations of thinhorn sheep. Some lived on snowy peaks near the huge glaciers and acquired a white pelage to match their surroundings, while the rest were confined to lower, forested slopes and became darker to help them blend in. Now that the glaciers have long since gone, the two subspecies interbreed freely in the intervening areas, producing sheep with a mix of their coloration. These hybrids typically have a greyish body but a white head and rump.

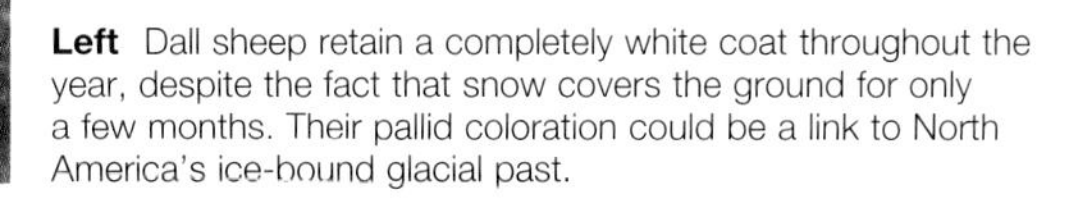

Left Dall sheep retain a completely white coat throughout the year, despite the fact that snow covers the ground for only a few months. Their pallid coloration could be a link to North America's ice-bound glacial past.

Bison

American bison helped to sustain the natural prairie landscape through their grazing and their never-ending search for fresh pasture. The indiscriminate slaughter of their herds ended this age-old link before it could even be studied. Today, most bison are domesticated and have lost the migratory urge of their ancestors.

Prior to the revolutionary wars, the Midwest was the Serengeti of the Americas. Its expansive vistas were dotted with plentiful herds of bison (also known as buffalo), together with pronghorns, deer, and *billions* of prairie dogs. These herbivores seemed to exist in limitless numbers – a recurring image in the creation myths of plains Native Americans was a hole in the ground from which bison bubbled forth like a freshwater spring. One herd of bison encountered by white settlers in the early 1800s was estimated to be at least four million strong.

MIGRATION PROFILE	
Scientific name	*Bison bison*
Migration	Historical mass migrations through American prairies
Journey length	Up to 320 km (200 miles) each year (present time)
Where to watch	Yellowstone National Park, Wyoming, USA
When to go	July–August

Above Bison grow a shaggy winter coat to keep out the bitter cold and can paw aside snow to expose the grass below. But when the snowpack is heavy, they must trek to sheltered valleys or face starvation.

UNSOLVED MYSTERIES

The commercial bison hunt of the mid-19th century was both devastating and rapid. Afterwards, Hunkpapa Sioux chief Sitting Bull declared that: "A cold wind blew across the prairie when the last buffalo fell – a death-wind for my people". The scale of the carnage has left us with little information about the species' natural history during its former prime. Where the bison travelled and how far, and how many there were, will never be known for certain.

At the start of the 19th century, the bison population was probably in the low tens of millions, found mainly in the shortgrass (western) and mixed-grass (central) prairie belts. These vegetation zones then covered around 1.5 million sq km (600,000 square miles), from the Rocky Mountains in the west, eastwards as far as a line from central Saskatchewan to Oklahoma, and south into Texas. In the more distant past, grassland flourished across most of what is now southern Canada, the USA, and northern Mexico, and so bison had the run of almost the entire continent apart from the far north. In summer the bison would have wandered through the open flatlands in huge herds, until winter snows forced them to disperse and take refuge in wooded river valleys or on lee slopes. It appears likely that they travelled up to several hundred kilometres a year.

New technology enables scientists to reconstruct lost bison migrations by analysing tooth enamel and bone collagen in old bison skeletons. As a bison grazes it acquires certain elements (especially carbon and oxygen) from its food, the concentration of which varies from place to place and according to climatic conditions. These biochemical signatures, called isotope ratios, help researchers to work out the animal's approximate location and when it lived, and what the climate was like. The technique can be used to study long-term ecological trends over many centuries.

DOMESTICATION

Bison and prairie alike are a shadow of their former selves. The prairies have been transformed into the breadbasket of the world, leaving only fragments of natural grassland unscathed, while 97 per cent of the bison alive today are undergoing domestication. Wild bison still exist, for example at Yellowstone and Wind Cave national parks in the USA and at Elk Island National Park in Canada, but they survive as islands in an ocean of domesticated relatives. The docile animals kept on ranches and private reserves, known as

Above Free-roaming bison survive in only a handful of reserves, and in numbers nowhere near those of two hundred years ago, but they are still thrilling to watch. A stampeding herd of bison is arguably the most iconic of all images of the American wilderness.

"bufftattle" or "beefalo", have mixed parentage. Genetically pure bison number fewer than 15,000. Of these, Yellowstone's bison are the sole truly free-range population.

The most visible bison migration in Yellowstone takes place each winter, when snow drives many of its herd downslope to sheltered valleys and forested areas. The size of each year's movement depends on the snowpack. In harsh winters, animals cross the park's northern and western borders, following the Madison and Yellowstone river valleys to reach lower-altitude wintering grounds, often following snow mobile tracks on their journey downhill. All this brings them into conflict with local ranchers, due to the small risk of bison transmitting brucellosis disease to their cattle, leading to controversial culls of bison that stray beyond agreed geographical limits. In the modern landscape, made up of parcels of land owned by different interests, the migratory behaviour of wild bison is a serious political problem.

PRESERVING WILDERNESS

Today there are ambitious plans to create huge protected bison reserves, by linking together existing national and state parks, preserves, and wilderness areas. The Buffalo Commons initiative, first mooted in 1987, envisages the removal of fences from the Great Plains in order to replace intensive cattle rearing with habitat restoration, so that the native herbivores can return to their original levels. Meanwhile, the Yellowstone to Yukon (Y2Y) scheme aims to safeguard a migratory corridor 3,200 km (2,000 miles) long in the Rockies.

RAILROAD TO RUIN

During the 19th century, hunting drove the bison almost to extinction within 30 years. Professional hunters shot thousands of bison at a time during the 1850s and 1860s, to feed labourers building North America's first transcontinental railroad. When complete in 1869, the railroad allowed bison carcasses to be swiftly shipped for processing. It also brought more hunters from the East Coast, accelerating the pace of killing through the 1870s. Hunters such as "Buffalo Bill" Cody became national heroes. The railroad industry actively persecuted bison, because migrating herds frequently blocked the line and presented a hazard to locomotives. Added to this wanton slaughter was the disruption caused by the spread of large-scale cattle ranching across the prairies. Bison found their traditional migration routes barred by corrals and barbed wire fences, and their pastures exhausted by farm livestock. By the mid-1880s, only a few hundred individuals were left from tens of millions.

Right Often bison were shot for their hide or tongue alone, or simply for sport. Their unwanted carcasses littered the prairies.

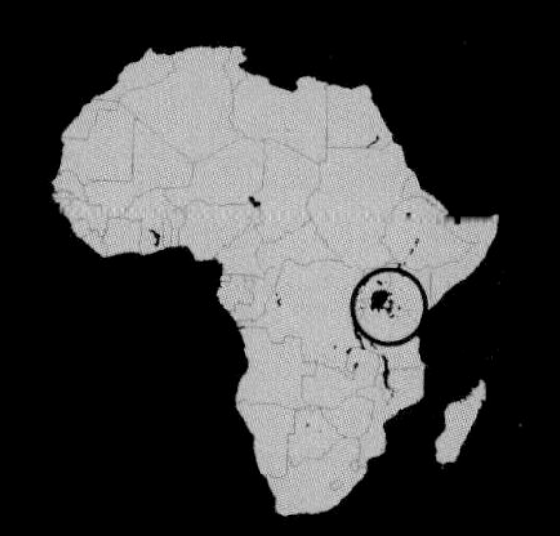

Sea of Grass

The African savanna is full of animal migrations, but for scale and drama none compares with the enormous herds of herbivores that thunder across the sunburnt Serengeti Plains, kicking up clouds of dust as they move in search of greener pastures.

Left Long lines of migrating wildebeest snake across Tanzania's Serengeti Plains. The great herds zigzag slowly but surely towards the horizon, often moving in single file.

Below, left to right Zebra are able to subsist on tall, dry, fibrous grasses, whereas wildebeest prefer shorter pasture; Thomson's gazelles are the fussiest eaters, preferring to crop the sweetest, juiciest shoots.

Savanna is a complex mosaic of open grassland, thickets, and woodland, shaped by unrelenting heat and an endless cycle of deluge and drought. The Serengeti–Masai Mara region, sandwiched between Lake Victoria and the Great Rift Valley escarpment, may be the best preserved example of this biome in all Africa. About 4 million years ago, it was the crucible for the evolution of bipedalism in humans. It also gave rise to an amazingly diverse community of migratory grazers, which provides food for many predators and scavengers, from lions and hyenas to vultures and dung beetles.

There are two wet seasons in the Serengeti–Masai Mara ecosystem. "Short" rains arrive in November–December, driving its migrant herds south; heavier "long" rains occur in March–May, and as they dry up signal the start of the return trek north and west. Numerically, the "big three" are wildebeest, Thomson's gazelle, and plains zebra. They coexist due to their staggered migration and selective feeding – a phenomenon known as grazing succession. Zebra set off first and crop the tallest, toughest, driest stalks, exposing the softer, more nutritious leaves and stems for the following wildebeest. The great piles of zebra and wildebeest dung produce a luxuriant lawn of fresh shoots that the gazelles, travelling last, nibble with their delicate mouths.

Other herbivores, including African buffalo and antelope such as impala, Grant's gazelle, hartebeest, and topi, stay in the same range all year, often in wooded areas. These species are less abundant. Only through migration can the big three maintain such large populations.

TOP HERBIVORES OF THE SERENGETI–MASAI MARA

SPECIES	BODY WEIGHT*	TYPICAL DIET	TYPICAL GROUP SIZE	MIGRATORY GROUP SIZE	POPULATION**
Wildebeest	165–290 kg (365–640 lb)	Lower stalks and leaves of average nutritional value	Herd of up to 500 females and juveniles; adult males separate	Tens of thousands	1.5–1.8 million
Thomson's gazelle	20–30 kg (45–65 lb)	Tender young shoots of highest quality	Herd of up to 150 females and juveniles; adult males separate	Several thousand	350,000–450,000
Plains zebra	220–320 kg (485–700 lb)	Tallest, coarse stems of often poor quality	Harem of 3–8 females and their young, led by stallion	Several hundred	200,000–250,000

* adult male ** in Serengeti–Masai Mara

Wildebeest

Wildebeest wandering across the plains of Kenya and Tanzania belong to a great migration involving more than two million grazing animals. These huge herds, which cultivate the grassland with their hooves, teeth, and dung, influence the shape of the whole savanna ecosystem.

The Serengeti and Masai Mara game reserves form a protected zone stretching either side of the Kenya–Tanzania border in Africa's Great Rift Valley and have been synonymous with animal migration since their fame spread around the world in the early 1950s. Their names are associated above all with the spectacular mass movements of common wildebeest, or white-bearded gnu, which track the seasonally changing distribution of rain-ripened grass through the region. Between 1.5 million and 1.8 million wildebeest roam the 24,000 sq km (15,000 square miles) of the Serengeti–Mara ecosystem, at certain times of year in breathtakingly large herds that stretch to the horizon. Migrating with them are up to 350,000 Thomson's gazelles and smaller numbers of eland and several other antelope species, together with about 200,000 plains zebras.

Wildebeest look ungainly and cattle-like compared to the sleek build of many of their relatives in the antelope subfamily (Antilopinae). They resemble an undersized bison, with long face, shaggy mane, chest tufts, and constantly flicking tail all creating the illusion of greater bulk. However, despite being somewhat dumpy, wildebeest make efficient migrants and have lots of stamina. By spending much of their life on the move, they are able to outmarch most predators, which ambush wildebeest herds passing through their home range but cannot keep up forever. Nomadism, in other words, is an effective strategy for minimizing the impact of predators that are not themselves nomadic.

CHASING RAINBOWS

Over thousands of years, wildebeest have evolved to exploit a food supply that is abundant (enabling them to live at very high densities) yet unstable (so they have to keep moving). Their migratory cycles follow the development of fresh grass in different places – they commute from one green island to another. It is as if they are chasing rain-bearing weather systems across the savanna. In fact, it has been suggested that their journeys may indeed partly be driven by changes in atmospheric pressure and humidity, to which grazers such as wildebeest are known to be sensitive, and by the visual stimulus of distant storm clouds.

The wildebeest of the Serengeti–Mara have a roughly circular migration route, covering up to 3,200 km (2,000 miles) a year in a clockwise direction. As elsewhere the timing and distance of this population's migration vary year by year, although two major movements stand out. From June to September, the wildebeest sweep northeast, with the main "push" usually in July–August; and from

AFRICA

WILDEBEEST RANGE

Total range

SERENGETI MIGRATION OF WILDEBEEST

Short grass plains

Woodlands

June

August

September–October

November–December

Serengeti National Park

KENYA

Kirawira

TANZANIA

Ngorongoro Conservation Area

MIGRATION PROFILE

Scientific name	*Connochaetes taurinus*
Migration	Annual circuit of Serengeti–Mara region
Journey length	Up to 3,200 km (2,000 miles) each year
Where to watch	Masai Mara National Reserve, Kenya
When to go	July–September

November until December, they drift gradually southeast. January to March is spent on their traditional calving grounds in the southern Serengeti, especially the Ngorongoro Conservation Area. Here the wildebeest pack into crowded assemblies that for a few weeks may reach 1,000 per sq km (2,500 per square mile). From the top of a kopje, or rocky outcrop, it is possible to see as many as 100,000 wildebeest in a single view. After calving has finished, the wildebeest move to the western portion of their migration loop, where the annual rut takes place in May–June, before swinging north again.

Wildebeest migrate at a steady plod, strung out over the landscape in many long, winding, single-file columns. They usually travel on a broad front along vaguely defined pathways – migrating herds are surprisingly tricky to locate without the help of a plane, and tourists often leave disappointed. At the swollen Mara River, however, wildebeest are forced to bunch together at a handful of crossings. There are only seven fording points, frequented by large numbers of Nile crocodiles. More wildebeest are probably crushed to death amid panic-stricken stampedes than are actually dispatched by the reptiles.

Below Occasionally, as in this aerial photograph, a wildebeest herd takes flight and hurtles across the savanna, but in general the migration proceeds at a gentle walking pace.

NORTH AND SOUTH

In most years the northern Serengeti–Mara receives twice the rainfall of areas further south. It is dominated by tall grass and scattered woodland, unlike the treeless short-grass country in the south. Why, then, do wildebeest abandon the more verdant north? The answer is that, even in these lush surroundings, the wildebeest herds quickly exhaust the land's resources. In addition, their southbound trek is also a quest for protein and minerals, especially phosphorus, which are scarce in the wooded northern pastures but plentiful in the sward sustained by the rich, volcanic earth of the south. Females particularly need these nutrients, to produce milk for their calves.

SAFETY IN NUMBERS

The wildebeest of the Serengeti–Mara produce almost half a million calves each year, at least 90 per cent of them within a two- or three-week window, sometime between the end of January and mid-March. The calves must get to their feet no more than 10 minutes after birth, since lions and spotted hyenas circle the herds, looking for easy pickings. There is a limit to the toll these carnivores can take, however, and the baby boom soon overwhelms them. The synchronized birth season is the result of an equally brief and intense mating season in late May or June, when hundreds of thousands of females go into oestrus at once. It is not known for certain what induces the females, dispersed over a wide area, suddenly to become receptive en masse. Perhaps the throaty rutting calls of the bulls – the loudest collective noise made by any gathering of herbivores on Earth – cause an "oestrus epidemic" to spread throughout the cows.

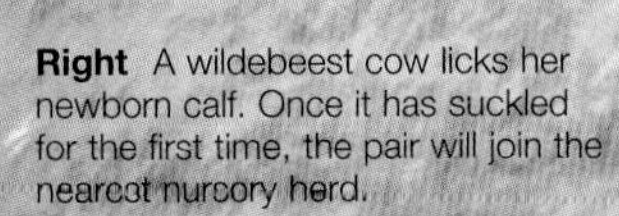

Right A wildebeest cow licks her newborn calf. Once it has suckled for the first time, the pair will join the nearest nursery herd.

African Elephant

For Africa's savanna elephants, life is one long migration, governed by the ebb and flow of rains across the continent. Close-knit family herds use their intimate knowledge of the land to locate water and essential minerals, and lone bulls make testosterone-charged journeys to track down receptive mates.

Elephants are among the very few terrestrial megafauna to have outlived the prehistorical era – others include rhinos, bison, and buffalo. They are heavyweight herbivores with appetites to match – a mature bull African elephant needs up to 300 kg (650 pounds) of food every day, equivalent to about 5 per cent of its body mass. The inevitable consequence is that these mammals need large home ranges and have developed a strongly migratory lifestyle.

The typical home range for a family of African savanna, or bush, elephants varies according to the availability of food and water. It could be about 15 sq km (6 square miles) in lush, high-rainfall areas or as much as 2,000 sq km (750 square miles) in arid regions, such as the Namib and Kalahari deserts. Elephants are also found in the rainforests of West and Central Africa, but the behaviour and migratory patterns of forest-dwelling elephants remain poorly known and are not discussed here (the map shows only the range of savanna elephants).

MIGRATION PROFILE	
Scientific name	*Loxodonta africana*
Migration	Seasonal movements to find water, minerals, or mates
Journey length	Each journey covers up to several hundred kilometres
Where to watch	Chobe National Park, Botswana
When to go	August–October

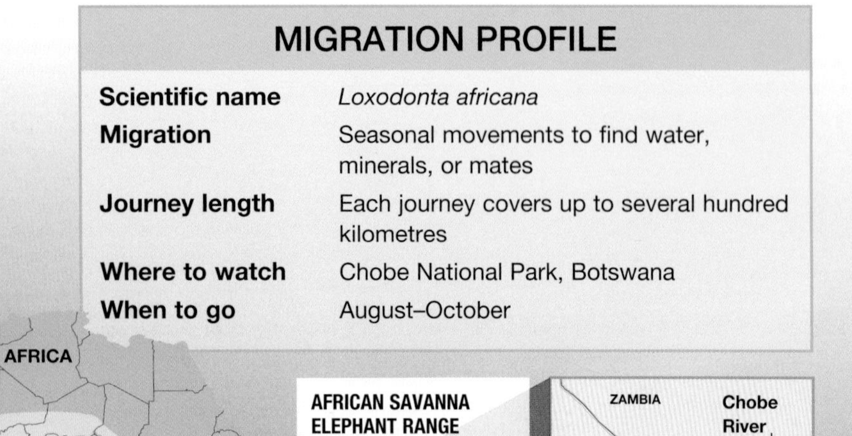

FEMALE LEADERS

In the savanna that covers much of East and southern Africa, elephants live in a female-dominated, or matriarchal, society. An average herd is made up of several adult females and their young calves, together with two generations of grown-up offspring. Members of each family unit develop extremely close, long-lasting emotional ties, and are led by one of the oldest, strongest females, known as the matriarch. Her leadership and years of experience are vital to the well-being of the herd; she is in effect the family historian, the guardian of a mental map full of information about water sources, seasonal food supplies, mineral-rich deposits, and dangers throughout their home range.

Savanna elephants live in a world of watery unpredictability. During the dry season, which can drag on for six or seven months, the few reliable water sources attract herds from far and wide. As the drought progressively worsens the elephants build up into spectacular concentrations, often in the company of other thirsty herbivores such as antelope, zebra, and African buffalo.

Each year, between April and October, the largest elephant gathering on Earth takes place along the meandering Chobe River in northern Botswana. By late August, when most waterholes in the area have long since dried up and midday temperatures hit 40°C (105°F), elephant numbers in Chobe will have reached about 45,000. Herds trek up to 325 km (200 miles) to visit the river, following familiar routes so that

Below Family groups of elephants travel in tight-knit formations, with their vulnerable young calves near the centre of the herd for protection.

huge avenues clearly visible from the air are scoured across the parched bush. Elephant "freeways" are used by a wide variety of animals and serve a useful function in savanna ecosystems, by turning over the soil and helping to promote plant growth.

MINERALS AND MUSTH

Several other factors, besides the need for fresh water and plentiful food, drive elephant migrations. Chief among them is dietary deficiency. Since grasses and foliage are often lacking in minerals, elephants must look elsewhere to find certain trace elements, especially iron, sodium, and phosphorus. Herds from the Kenya–Uganda border area travel to the slopes of Mount Elgon, an ancient volcano, where they enter caves to scrape away the abundant mineral salts. Generations of elephants will have made an identical journey to Mount Elgon; knowledge of the caves and how to get there is an important part of their inherited culture, passed from mother to daughter.

Since older male elephants lead a solitary existence, they must undertake periodic breeding migrations to find partners. Their journeys are triggered by a rush of reproductive hormones, particularly testosterone, which can soar to 50 times normal levels. A bull in peak breeding condition, said to be in musth, becomes restless and aggressive, and in his frenzied, hormone-fuelled state may wander several hundred miles within a month. Bulls usually hunt for mates at the onset of the dry season, having spent the wet months fattening up, although this migratory behaviour, like female oestrus, can occur at any time of year.

LONG-RANGE COMMUNICATION

It seems that bulls in musth pinpoint female herds by listening out for the super low-frequency contact calls – a form of infrasound – that groups of elephants use to stay in touch. These powerful rumbles carry amazingly long distances, especially in cool, still air, which is why elephants vocalize mostly at dusk and just before dawn. The sound waves are transmitted by the ground itself, too. As a result, elephants can detect them with the nerve-packed vibration sensors in their feet, called Pacinian corpuscles.

There is growing evidence of the pivotal role played by sound in elephant migration. Herds may even be able to sense distant thunderstorms from up to 250 km (150 miles) away, enabling them to head directly towards the rains and the promise of fresh grass.

Left Elephants rampaging through villages and farms is becoming a more frequent occurrence in southern and eastern Africa as people increasingly encroach on land that was formerly open and unfenced. Balancing the conflicting needs of the local population and migratory elephants is a growing problem.

BARRIERS TO MIGRATION

Over thousands of years savanna elephants evolved migratory patterns to cope with periods of drought, but in the space of the last hundred or so years their homeland has experienced radical change. Vast swathes of sub-Saharan Africa are now a patchwork of fenced reserves, game lodges, and farms, making it increasingly difficult for elephants to move freely. Many herds are effectively corralled in small areas. When protected from poachers, their numbers can exceed the carrying capacity of the habitat, and the over-abundant elephants begin to destroy woodland areas by uprooting trees and trampling plants. One solution is to cull elephants to keep their populations within sustainable limits. From 1966 until 1994, this was the strategy adopted in South Africa's Kruger National Park, where the optimum elephant population was estimated to be 7,000–7,500. Since the cull ended in 1994, some of the park's fences have been torn down to encourage the elephant population to regulate itself through emigration. Another non-lethal management technique is to move surplus elephants to entirely different areas.

From Desert to Delta

Southern Africa's fertile Okavango Delta is a world away from the parched plains, salt pans, and sandy desert that press in from all sides. When its annual floodwaters arrive, the emerald-green and sapphire-blue wetland offers a temporary sanctuary for massive concentrations of wildlife.

In April and May the Okavango Delta bursts into life, transforming the northwest corner of Botswana into one of the largest wetlands on Earth. It is fed by torrential rains in distant Angolan mountains, which create a pulse of water that due to a quirk of geology advances inland, instead of towards the sea. The surge carries 10 billion m (350 billion cubic feet) of silt-laden water down the meandering Okavango River to replenish a maze of low-lying lagoons, swamps, and floodplains. Eventually the marshy wilderness spans 15,000 sq km (5,800 square miles), but by September it is already shrinking because of rapid evaporation in the fierce heat.

The inundation triggers a huge influx of herbivores, especially elephants, plains zebra, buffalo, wildebeest, hartebeest, and springbok. Most come from the northern Kalahari Desert and the Makgadikgadi Pans – a vast area of salt flats that contains water from November to April but turns into a lifeless thirstland at other times. Once in the haven of the delta, the migrant herds often follow the well-worn channels made by local hippos. Some stay here throughout the Kalahari's dry season, while others push hundreds of miles further north, as far as the Linyati Swamp and Chobe River, before finally drifting south again.

Above This satellite photograph, taken from the north, shows the arid plains of Botswana, with the distinctive "panhandle" form of the Okavango Delta clearly visible.

Left Elephants wade through one of the delta's myriad water channels after the return of the winter floods. Proficient swimmers despite their size, they use their trunks as snorkels to cross deep lagoons.

Below, left to right The floodwaters trigger explosive vegetation growth, ensuring a plentiful supply of food for the herbivores that converge on the delta from far and wide. Species on the move include buffalo and zebra, which trample across the verdant pastures and swamps in large, mobile herds, closely followed by predators such as lions.

Mongolian Gazelle

More than a million gazelles roam the eastern Mongolian steppes, one of the world's largest surviving grassland ecosystems. The restless herds seldom stay still for long and trek thousands of miles a year, but the complex movements of these jittery, highly strung nomads are not fully understood.

We tend to associate the sight of massed ranks of grazing animals with the East African savanna, or with the American Midwest in its pristine state before the arrival of European settlers. Compared to these famous wildlife spectacles, the huge gatherings of gazelles on the steppes of eastern Mongolia are virtually unknown. The reason is simple. The isolation of the steppes and their punishing climate of bitterly cold subarctic winters and scorching hot summers, when daytime temperatures soar to 40°C (105°F), make study of the Mongolian gazelles' life-cycle a daunting proposition. Searching for such wary, mobile animals in an endless sea of grass, covering an area of 260,000 sq km (100,000 square miles), is like hunting the proverbial needle in a haystack.

UNTAMED WILDERNESS

Visitors to the steppes often comment on how barren and deserted they appear. Devoid of trees, hedges, or roads, these lands are seemingly featureless: a lonely, rolling plain stretching to the horizon in all directions. But in fact, this is a highly productive biome, capable of supporting large concentrations of wild ungulates, or hoofed mammals. In 1989, the population of gazelles in eastern Mongolia and neighbouring regions of southern Russia and northeast China was estimated at one to two million, with smaller numbers found in western Mongolia.

The steppes in Mongolia's remote east are intact across wide areas, unlike the American prairies and many other temperate grasslands around the world that survive only as fragments. The natural steppe vegetation, dominated by feather-grasses with a

Above Highly mobile groups of gazelles melt into the shimmering haze during the fierce midday heat, but are easier to spot in the low light of dawn and dusk.

Facing page Sharing the steppe with the gazelles are nomadic pastoralists, who travel on horseback and camp in round tents called yurts. Increasingly their livestock compete with the wild herbivores for grazing and precious water supplies.

scattering of dwarf shrubs, has survived due to the scarcity of surface water, which limits the opportunities for raising cattle and sheep and means that arable farming is next to impossible.

TRUE NOMADS

Mongolian gazelles are well adapted to this tough environment. They can cope with prolonged drought, winter cold, and summer heat – and above all, do not linger. One study found that even outside their peak migration periods the gazelles move at least 19 km (12 miles) a day. Their nomadic lifestyle is driven by a constant quest for fresh pasture and the need to escape thick snow, which in some areas blankets the ground until April. Contributory factors might include the search for essential mineral salts and the avoidance of predators and swarms of biting insects.

In common with most antelopes, Mongolian gazelles keep out of danger through impressive speed and by living in groups for extra vigilance against their enemies – mainly wolves. They are said to be able to sprint at 65 kph (40 mph) over distances of as much as 14 km (9 miles). A typical group comprises 20–30 gazelles, and may swell to 100–120 individuals during the winter. In spring and early summer the groups briefly become much larger and extremely fluid: different herds merge and break up, sometimes forming loose aggregations

VALUABLE INSIGHTS

Mongolian gazelles face a growing number of threats, including hunting, oil drilling, grazing competition from cattle, and artificial barriers on their migration routes, such as fences and pipelines. To help develop a conservation plan for the gazelles, the Wildlife Conservation Society (WCS) of North America has launched a research project into their life-cycle and ecological needs. The researchers capture newborn calves, weigh them, check their health, and establish the sex ratio of each new generation. They also test for illnesses such as foot and mouth disease, which can spread between the gazelles and domestic cattle, leading to conflict with livestock owners. Further surveys of the region have provided the first accurate data for the size and structure of Mongolian gazelle populations.

Right WCS field workers check the weight of a young calf.

MIGRATION PROFILE

Scientific name	*Procapra gutturosa*
Migration	Year-round nomadic movements
Journey length	Each movement covers up to several hundred kilometres
Where to watch	Choibalsan area, eastern Mongolia
When to go	June–July

6,000–8,000 strong. The scientist George Schaller has described how on occasion several tens of thousands of gazelles form a superherd for a single evening, then disperse again after dawn.

SYNCHRONIZED CALVING

Mongolian gazelles stage their rut in November and December. As the snows melt in spring, the herds can move northwards across the steppes towards their summer feeding areas, travelling up to 300 km (180 miles) per day and swimming across rivers that block their path. In late June, having arrived at their traditional calving sites, the heavily pregnant females come together to give birth.

How the gazelles navigate to their calving grounds is not known, but for a short while these places host the biggest assemblies of large mammals anywhere in Asia. Male gazelles gather near the edge of the calving areas, leaving the females massed in single-sex herds. Each mother produces one or two babies and two-thirds of all the females give birth within a week of each other, so the grassland is soon dotted with young gazelles. By a week old, the youngsters follow their mothers on short foraging trips; after a couple of weeks or so, the gazelles end their temporary sedentary existence on the calving grounds to return to a life of wandering.

Synchronized breeding is a predator-beating strategy that these gazelles share with other plains animals, such as caribou, wildebeest, and snow geese (*see* pages 44–45, 56–57, and 118–119). By producing a sudden glut of young, these species overwhelm their predators with potential prey, allowing a greater proportion of their offspring to survive due to limits on the amount of prey the predators can consume.

Norway Lemming

Norway lemmings are record-breaking breeders, and in some years their numbers skyrocket. Huge armies of the starving rodents erupt from Scandinavia's mountains to swarm across the lowlands in a frantic search for food, fighting among themselves and destroying crops.

Rodents make up over 40 per cent of the world's mammal species, yet few of them can be said to be truly migratory – their small size prohibits regular long-distance travel, and most spend their entire life in the same territory. Lemmings are without doubt the most famous rodent migrants. Their occasional population booms and subsequent invasion of neighbouring areas have fascinated people for centuries, but understanding what lies behind these phenomena is a classic scientific problem. Research into the dramatic fluctuations in lemming numbers from year to year, known as lemming cycles, has so far failed to produce a definitive explanation.

All five species of lemmings in the genus *Lemmus* have "cyclic" populations and live in the Arctic. In normal years, Norway lemmings are restricted to northern parts of Norway, Sweden, and Finland, and the northwestern corner of Russia. Like their relatives, they have dense, velvety fur and hairy feet for protection against the cold, but whereas the other species inhabit open tundra, these lemmings prefer damp birch and willow woodland in highland regions.

PROLIFIC REPRODUCTION

Expert burrowers, Norway lemmings dig a system of tunnels as a summer home, emerging round the clock (there is permanent daylight this far north) to crop the lush carpet of grasses, sedges, and herbs on the woodland floor. In winter, they abandon their warrens and move uphill to snow-covered peat bogs, where they clear runways beneath the blanket of snow. Safe in these frozen passages, the lemmings do not hibernate but remain active all winter, nibbling roots and mosses to survive. Incredibly, females can breed throughout the winter.

Female lemmings outnumber males and have phenomenal reproductive potential. During the spring and summer, when food is abundant, each female produces litters of up to 12 young after a gestation of only 16–20 days, at intervals of about a month; meanwhile, her female offspring all start breeding by the time they are two or three weeks old. As a result, lemming populations soar if conditions allow, sometimes increasing by a factor of 200 between spring and autumn. When this happens, the animals become so numerous that hill-walkers can scarcely put a foot down without scattering them in all directions.

Below Lemmings are surprisingly adept swimmers. Overcrowding can get so severe during population booms that famished lemmings will cross lakes, rivers, or even coastal bays in a desperate bid to track down food.

Above When lemming numbers peak, predators such as this Arctic fox enjoy easy pickings, the sudden glut of food enabling them to raise large families.

Population growth on such a scale cannot continue unchecked, and so after three to five years the lemming numbers crash suddenly due to starvation and lack of space. The cycle repeats over and over again, with a particularly high peak once every 30–35 years, when chronic overcrowding triggers a mass migration. A tide of lemmings now surges across country at up to 16 km (10 miles) a day.

REMOVAL MIGRATION

Migrating lemmings do not orientate in a specific direction. Driven by blind hunger, the refugees disperse randomly and cross all manner of barriers, from rivers to lakes and roads, often suffering high mortality; in the 1970 lemming migration, for example, 20,000 were squashed along a 195-km (120-mile) length of road. The largest concentrations occur where the terrain forces them to bunch together, and in these bottlenecks the stressed lemmings become aggressive to each other, as well as quite unafraid of humans. Very rarely, these animals will tumble over cliffs or attempt to swim out to sea.

Scientists class these spectacular events as removal migrations – one-way movements carried out with no intention of returning, rather like human emigration. Besides lemmings, certain other rodents from northern latitudes undertake removal migrations, including the Scandinavian field vole, the Japanese grey-sided vole, and the Alaskan meadow vole. In the long term, such drastic behaviour enables these species to expand their range by colonizing new territory, with the result that they may be able, in theory at least, to adapt to gradual changes in the climate.

MIGRATION PROFILE

Scientific name	*Lemmus lemmus*
Migration	Population explosion forces lemmings on the march
Journey length	Up to 160 km (100 miles)
Where to watch	Highland woods and bogs in northern Scandinavia
When to go	May–August (migrations are irregular and occur in autumn)

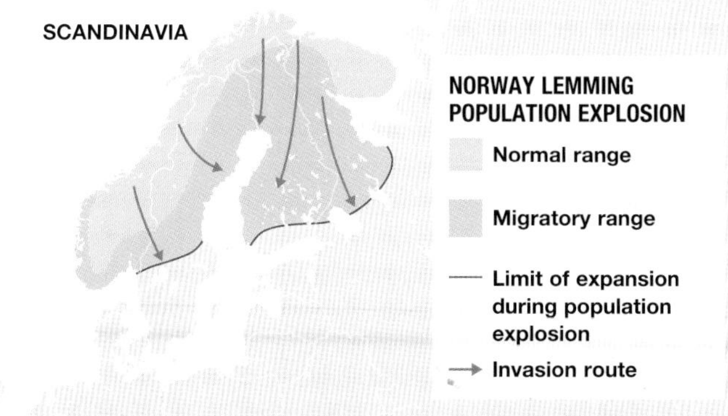

LEMMING LEGENDS

A wealth of folklore has been inspired by "lemming years", some of it dating back to the Middle Ages. The oldest recorded legend, from the mid-16th century, described how swarms of lemmings formed in the clouds and fell to Earth like rain, and priests kept records of lemming migrations because they were considered a sign of impending war. Inuit peoples passed on a variety of stories about lemmings, including that they emerged from snowstorms and were creatures from outer space. Early zoologists came up with the idea that lemmings swam out to sea to rediscover their ancient homeland, a kind of Atlantis for rodents. But the most persistent myth is that hordes of lemmings commit suicide by rushing headlong off cliffs to drown in the sea. The 1958 Disney documentary *White Wilderness* notoriously included fake footage of a mass suicide, staged using captured animals, although there have been serious attempts to rationalize the behaviour, by arguing that the maddened lemmings may have eaten toxic plants after exhausting their usual food supply.

Right In invasion years, packs of lemmings turn up in urban areas, making an incongruous sight in parks and gardens. Many of the migrants end up as roadkill.

Emperor Penguin

The gruelling marches made by emperor penguins to and from their breeding colonies on the Antarctic ice shelf have come to symbolize the perilous nature of life on the "Great White Continent". Parent penguins take turns to trudge across the frozen wastes to reach their waiting chick, braving some of the harshest weather on Earth.

Most animals that frequent the Southern Ocean's pack-ice zone, among them the humpback whale and Arctic tern (*see* pages 78–79 and 142–143), are long-distance migrants that visit for only for a couple of months in summer, then head north again. Very few species live here all year round. Without doubt the best-known Antarctic resident is the emperor penguin, which has a circumpolar range that encircles the continent. It seldom ventures into the warmer waters that lie beyond, its life-cycle being ruled by the seasonal ebb and flow of ice.

WORLD OF ICE

Between January and March – late summer in the Southern Ocean – emperor penguins range through the cold, nutrient-rich waters that wash the shores of Antarctica and its outlying islands. The penguins hunt krill, fish, and squid, following their prey to the most productive parts of the ocean. When the sea begins to freeze over in March, the penguins of breeding age (four years old and over) head south and congregate offshore, waiting until the ice is sufficiently thick to support them. This is the cue for them to haul out and start their trek inland.

Few emperor penguins actually set foot on dry land. Instead they travel across "fast ice": sheets of floating ice that have fastened to coasts. Of the 40 known emperor breeding colonies, nearly all are on fast ice. The ice must remain anchored until the end of December for the penguins to breed successfully; if it is too thin and breaks up early, the chicks will probably be cut off from their parents or swept to their deaths. This means that the length of the penguins' migration varies. In areas with thick, strong ice their colonies are only several miles from open water, whereas colonies may be up to 200 km (125 miles) from the ice edge if the ice is less stable.

THE LONG MARCH

In March and April the penguins shuffle to their colonies, forming long lines that snake across the vastness of the ice like ant trails. Clearly they cannot navigate to the right spot using visual landmarks, because the icescape is constantly shifting. One theory is that they use the reflection of water on clouds, which creates a form of "water sky", but this can explain only how they navigate towards the sea, not away from it.

Once at the breeding area the penguins quickly pair up and the females lay a single egg after a noisy courtship, usually in May or early June. The females promptly return to the sea to feed, leaving the increasingly hungry males to incubate the eggs through the perpetual darkness of the Antarctic winter. During the nine-week incubation, the male penguins huddle in dense crowds, in the lee of an ice shelf where possible, to withstand howling winds that push the chill factor to –60°C (–75°F). They have pretty much the whole continent to themselves, since no other vertebrates apart from seals and humans overwinter on Antarctica.

Below Parent emperor penguins must waddle across sheets of coastal ice to fetch food for their chick. Adapted for swimming rather than moving on land, they can manage no more than an awkward shuffle and often toboggan on their bellies to save energy.

Above A lone adult supervises a crèche of half-grown young while the colony's other penguins are away fishing. The youngsters press tightly together for warmth, facing into the huddle with their heads lowered most of the time.

PERFECT TIMING

In July the female penguins walk back to the colony, having fattened up during several weeks of intensive fishing. They time their arrival to coincide with the hatching of the colony's chicks. Even if they arrive a week late, all is not lost, because the males, despite being half-starved, are able to secrete a milky fluid from their oesophagus to give the chicks a life-saving drink.

Presumably each female locates her partner among the scrum of huddled males by listening for his unique call, and after greeting him she gives the chick its first proper meal: a protein-packed slurry regurgitated from her crop (a food storage area in the throat). At last the male is relieved of his duties and hurries to the sea to break his fast. This is arguably the most impressive migratory feat in emperor penguins, as a typical male will not have eaten for 13–16 weeks and loses 40–45 per cent of his body weight in the process.

For six weeks the parents operate a rota to the sea, returning each time with a cropful of food for their chick, and from then on it is old enough to be left in a crèche with the rest of the colony's young while both adults go fishing. As the ice melts the foraging expeditions get shorter, so the wait between feeds is shorter too. On average an emperor penguin chick will be fed a dozen times before it is ready to undertake its own migration to the sea in December or early January. The species' entire breeding cycle is calibrated so that the five-month-old chick arrives in summer when food is plentiful and it has the best chance of survival.

MIGRATION PROFILE

Scientific name	*Aptenodytes forsteri*
Migration	Commutes between breeding colonies and the sea
Journey length	Up to 200 km (125 miles) per trip
Where to watch	Snow Hill Island, Weddell Sea
When to go	November–December

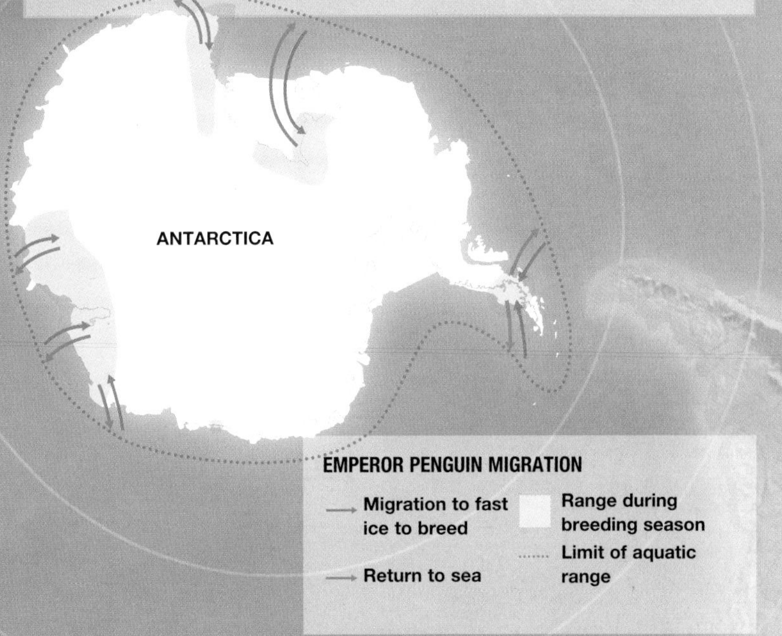

CLIMATE CHANGE THREAT

The emperor penguin depends on its ability to transport food and fast for long periods, but its finely balanced migrations are made more hazardous by global warming. In the 1990s French biologists examined 40 years' worth of emperor population data from the colony at Point Géologie. In years with higher sea temperatures and less ice, the birds travelled further to breed and their krill food source was scarcer, resulting in higher death rates and lower birth rates. During a prolonged warm period in the late 1970s, the Point Géologie colony shrank by 50 per cent.

Red-sided Garter Snake

For a few weeks in spring the planet's largest gathering of snakes takes place on the marshy plains of northwest Canada as writhing masses of red-sided garter snakes emerge from hibernation. These dramatic scenes return in the autumn as the snakes pour back into their dens to sleep out the coming winter.

MIGRATION PROFILE

Scientific name	*Thamnophis sirtalis parietalis*
Migration	To and from hibernation dens
Journey length	Up to 20 km (12.5 miles) each way, possibly more
Where to watch	Narcisse Wildlife Management Area, Manitoba, Canada
When to go	Late April–May

Red-sided garter snakes slip into a deep sleep for the coldest eight months of the year, then awake to spend their brief summer in a frenzy of breeding and feeding. This extraordinary stop-start existence is the only way that these cold-blooded animals are able to survive the extreme climate of central Canada, where temperatures plunge to –40°F (–40°C) in winter. A cycle of prolonged torpor broken by bursts of activity requires accurate time-keeping and places tremendous demands on the garter snakes' body processes, which slow right down until only the faintest flicker of life is left.

LIFE IN THE FREEZER

Red-sided garter snakes are a subspecies of the common garter snake, a species common in much of North America. Of the 11 different subspecies, they have easily the most northerly distribution: the northern limit of their range extends from British Columbia, east through Northwest Territories to Ontario, reaching up to 60°N in places. No other American snake occurs this near the Arctic. Garter snakes produce a very mild venom, but are not dangerous to humans.

Although mass hibernation is far from unique to red-sided garters, they can muster the most spectacular aggregations. Each autumn at the Narcisse Wildlife Management Area in Manitoba, an estimated 50,000 of them spill into four major wintering dens and several smaller dens. The size of these gatherings is caused by two key environmental factors coming together. First, there is a lot of ideal garter snake habitat (ponds and swamps) at Narcisse, with an abundance of their favourite prey (frogs), giving rise to a big snake population. Second, the area has high-quality hibernation sites (deep crevices and caverns in limestone outcrops), which encourages the snakes to migrate to these favoured locations in huge numbers.

Herpetologists conducted an experiment to investigate the freezing tolerance of red-sided garter snakes and established that in autumn they can survive for a brief spell even if their body fluids become 40 per cent ice. However, the snakes are likely to die after about 10 hours and lose this ability altogether in the depths of winter – in other words, their freezing tolerance is a short-term tactic for

Left A warm, sunny day in spring prompts garter snakes to erupt from their underground hibernation dens. For several days the rocky ground is transformed into a seething cauldron of reptilian bodies.

Above Canada's swamps and reedy pools provide ideal frog-hunting habitat for red-sided garter snakes.

coping with unseasonally early cold snaps in autumn. It cannot see them through an entire winter, and this is why they must head down into the ground to seek refuge below the frost line. Once inside their den, or hibernaculum, the snakes can safely "shut down" and soon carpet every available rocky surface in immobile heaps.

MATING BALLS

When the air temperature reaches about 25°C (77°F), usually in mid-April, the snakes become active again. Males leave the den first and crowd around the entrance to mate with the late-emerging females, getting warmer and more energetic while they wait. The female snakes are much thicker and longer, and heavily outnumbered by the males that fight to copulate with them. In their effort to achieve a mating, the wrestling males twist themselves into wriggling balls. This annual orgy lasts a few days or may erupt several times over a period of up to three weeks, depending on the weather, after which the snakes disperse into the surrounding area to hunt for food and, in the case of females, give birth. Each female produces 10–50 live young. The snakelings will spend their first winter hibernating in widely scattered burrows or ant hills, and do not migrate to traditional dens until their second year.

HOME SWEET HOME

The homing mechanisms of snakes are poorly understood, but their ability to detect tiny traces of chemicals – the main sense in snakes – has long been assumed to play a part in their migration. A snake's forked tongue constantly picks up scent molecules from the ground and air, passing them to supersensitive taste receptors, called Jacobsen's organs, in the roof of its mouth. In this way the snake builds up a detailed scent map of its surroundings, including the distinctive odour of its denning area.

Snakes are also believed to use thermal clues, and researchers at Oregon State University are investigating if garter snakes are able to be guided by localized variations in the Earth's magnetic field. Some of the red-sided garters at the Narcisse snake dens have been marked and recaptured by researchers, proving that the majority return to the same location each year, although a few snakes were found to rotate between different dens. Many other questions still remain unanswered. It is not yet known exactly where the Narcisse snakes give birth and spend the summer, nor whether they follow the same routes to and from their dens or simply disperse and return from all directions.

SNAKE MIGRATIONS

Garter snakes are not the only snakes to migrate: many other species in temperate regions converge on shared dens to escape winter cold, the reptilian equivalent of bats or insects hibernating in groups. North American snakes that do this include black rat snakes, prairie and western diamondback rattlesnakes, and gopher snakes. Sometimes different species overwinter together, especially in places where suitable hibernation sites are few and far between. In Scandinavia, the adder or common viper also assembles at winter dens – a strategy that enables it to live in open tundra north of the Arctic Circle, earning this species the title of northernmost snake on Earth. There is little data on the distances individual snakes travel to reach their hibernation sites, but they are thought to range from hundreds of yards to several miles.

Above The garter snake's snout contains sophisticated sense receptors, which it uses to analyse scents that aid it in migration.

Galapagos Land Iguana

Known by early seafarers as "dragons", these large lizards are unique to the Galapagos islands in the Pacific Ocean. The female iguanas on one island climb to the top of an active volcano to lay their eggs in its hot ash – perhaps the most unusual migration of any reptile.

MIGRATION PROFILE

Scientific name	*Conolophus subcristatus*
Migration	Females climb La Cumbre volcano on Fernandina Island
Journey length	Up to 16 km (10 miles) each way
Where to watch	Galapagos islands, East Pacific
When to go	June–July

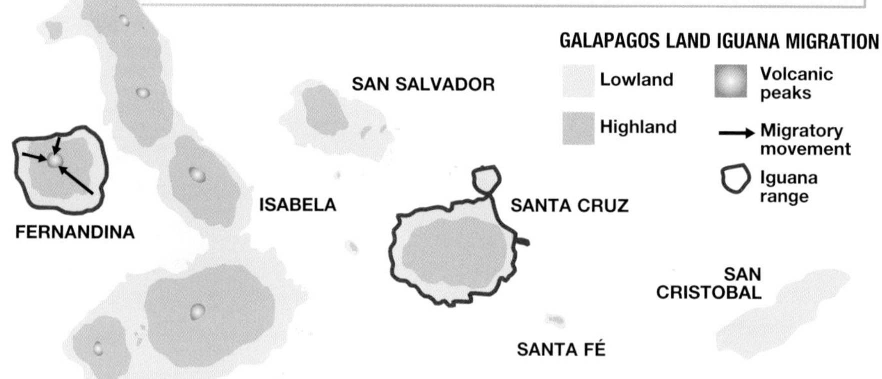

Above Having struggled to the summit of Volcán La Cumbre, a female iguana immediately begins digging a burrow for her eggs. Her powerful hind feet and claws make quick work of the loose volcanic ash.

With their thickset jaws, spiky crests, and heavy, wrinkled bodies, land iguanas bear a striking resemblance to the popular image of what a prehistoric creature would look like. Their long, sharp claws and large size – some males grow to 1.2 m (4 feet) long and weigh up to 12.5 kg (28 pounds) – enhance this impression, so perhaps "dragon" is not such a far-fetched name after all. Today there are two known species of land iguana – *Conolophus pallidus* is found only on the island of Sante Fé, while *C. subcristatus* lives on six islands in the west and centre of the Galapagos archipelago.

CACTUS EATERS

Land iguanas shun the greenest parts of their island homes and instead favour lava fields and areas of scrub in the driest and most volcanically active zones. Their yellow, rust, or pinkish skin provides a striking contrast to the monotonous grey and black moonscape. Few plants grow here, yet the adult iguanas manage to survive on a chiefly vegetarian diet consisting of the juicy pads, fruits, and flowers of the opuntia cactus, which also supplies all the water they need. In the absence of large land mammals, these lizards have, together with the islands' famous giant tortoises, become the main herbivores in the Galapagos.

In May the hot season is nearly over on Fernandina, at the western edge of the Galapagos island group, and its land iguanas gather to breed. Fully grown males, which reach sexual maturity at 8–15 years old, each defend a small territory, to which they try to tempt females. The most brightly coloured males with prime real estate assemble the largest harems, but weaker males sometimes achieve "sneaky matings" on the fringes of the mating ground.

VOLCANO JOURNEY

In June the weather suddenly changes, as a drizzle-laden mist called the garúa rolls in from the sea. This may be one of the cues that drives the pregnant female iguanas to set off on their arduous migration across the unforgiving terrain of the lava fields and up the steep slopes of Volcán La Cumbre that rise to 1,460 m (4,800 feet) in the middle of the island. They travel furthest on cool, foggy days, pausing to rest when the garúa clears and the temperature soars.

The females take up to 10 days to haul their bodies, by now heavy with eggs, to the rim of the volcano's crater. Some lay their eggs here, fighting over the best positions, and the rest descend into the volcano itself by scrambling down the unstable scree slope to the

SUN WORSHIPPERS

The lives of land iguanas are governed by the sun. They emerge cold from their burrrows, crawling sluggishly into the first rays of morning sunshine. They bask with their bellies facing the sun for half an hour, steadily heating their bodies to the critical level of around 37°C (99°F) that will enable them to begin foraging. They shelter from the midday heat in the shade of cacti or rocks, but as the midday heat melts away there is a second peak of lizard activity. In late afternoon they return to their burrows to conserve as much body heat as possible during the cold night.

Above Land iguanas frequently bask on sun-warmed rocks on the arid plateau of their volcanic island homes. Prickly pear cacti provide the adult iguanas with almost all of their dietary needs.

Left The iguanas possess a formidable armour of large, overlapping scales. When excited, the lizards' head, crest, and flanks flush with heightened colour.

caldera floor, a tiring journey of about 1 km (just over half a mile). The reason for the iguanas' trek now becomes apparent – they have come to lay their eggs in volcanic ash kept warm and slightly moist by the hot sulphurous gases and steam escaping from fumaroles (vents in the volcano). The centrally heated ash is a perfect incubator.

A female iguana needs several days to dig her nest, lay up to 20 or so eggs inside, and re-cover the entrance. Hiding the tunnel is vital because suitable nest sites are in demand – other females might try to save time by excavating the same spot, trampling the eggs. As an extra precaution, many females guard their nest for a few days before leaving. The nesting season on Fernandina typically lasts six weeks, during which several thousand female iguanas make the month-long round trip up and down the volcano. On islands without an active volcano, the iguanas carry out much shorter migrations and nest in the nearest area of damp sand they can find.

BORN OF FIRE

Land iguana eggs hatch after 85–110 days and the babies take about a week to scrabble out of their tunnel. Once in the open, the young iguanas are helped by a combination of speed and their speckled camouflage, but many still fall victim to aerial strikes by Galapagos hawks or ambush by snakes.

How land iguanas came to be on the Galapagos islands in the first place, which are both very young in geological terms and extremely remote, lying 960 km (600 miles) off the coast of Ecuador, has long been the subject of speculation. The scientific consensus is that they are descended from iguanas that survived an epic one-way voyage from Central or South America. Thousands of years ago these castaways, or their eggs, must have reached the islands after drifting across the Pacific on a raft of logs or floating vegetation.

Above For male toads, the urge to mate becomes all-consuming in spring. A willing partner is gripped tightly to stop rivals prising her away.

MIGRATION PROFILE

Scientific name	*Bufo bufo*
Migration	To and from freshwater breeding site
Journey length	Up to 2.5 km (1.5 miles) each way
Where to watch	Ponds and creeks throughout Europe
When to go	February–April

TOAD ROADKILL

Many toads never complete their spring migration, having met an untimely death on a busy road. Dazzled by bright headlights, these sluggish animals are unable to dodge oncoming vehicles and get squashed in their thousands. To make matters worse, the toads often have to cross several roads to reach their breeding pond, and because they travel at dusk in early spring, their migration coincides with heavy rush-hour traffic. Every year an estimated 20 tonnes of toads are killed on British roads alone, and in some places the high mortality rate may cause the extinction of the local breeding population. To reduce this death toll volunteers organize patrols to help the toads across roads at peak migration times and temporary warning signs are erected to alert drivers. These measures are not needed for the toads' return journey as they disperse to their feeding areas in smaller groups later at night, when roads are quieter.

European Common Toad

Driven by blind instinct, European common toads crawl to their spawning ponds on mild, wet nights in early spring. Most head for the same pond within a few days of each other, advancing in large groups like nocturnal armies on the march. Soon the water is teeming with amorous amphibians frantically competing to get a mate.

Common toads must keep their permeable, oxygen-breathing skin moist, and need fresh water for their eggs and tadpoles to develop, yet they spend nearly all of their adult life on land. This paradoxical existence is possible only because they are active in the cool of night and migrate to wetlands to breed – two key behavioural strategies adopted by frogs and toads the world over.

Favourite haunts of common toads include rough grassland, scrub, woodland, and gardens, where they spend the day hidden in damp corners under stones or logs. The toads are well adapted for life on land: their hind feet are only partly webbed and their front feet have lost webs altogether, while their strong legs are ideal for burrowing – and for shovelling spiders, earthworms, and slugs into their mouth. These are voracious predators that will devour any animal they can swallow whole. But they do not feed in water, so the aquatic phase of their annual life-cycle is necessarily brief and has a single purpose: reproduction.

BROKEN MIGRATION

Shortly after nightfall sometime between February and April, depending on the latitude and weather, groups of common toads travel together to their breeding site, marching with a wide-legged gait that looks fairly comical to our eyes. This is the final stage of a migratory journey that began months earlier. Having spent the previous summer dispersed across their foraging grounds, the toads first started to move back towards their spawning site in August or September. This was a gradual, indirect movement, covering perhaps several hundred metres or a kilometre or two. The first cold snap of autumn stopped the toads in their tracks and stimulated them to enter hibernation in abandoned rodent burrows and piles of leaf litter.

The migration is restarted by rising temperatures and, to some extent, rainfall. It is speculated that during the later stages of their hibernation, the toads' biological clock switches on a mechanism responsive to changes in the temperature of the soil around them. As the ground warms up, the toads stir into life. However, they do not set off until the air temperature at dusk reaches about 5–6°C (41–43°F) for several days, and the toads also wait for a rainy night. In mild winters, the migration gets underway exceptionally early; conversely, it is delayed by prolonged cold weather.

HOMING INSTINCT

Consumed by the urge to breed, common toads display an astonishing determination to reach their breeding pond or creek, crossing obstacles such as roads and railroad tracks and scrambling over walls. One study in Germany found that the toads travel an average of 50 m (160 feet) a night, sticking to a more or less straight course. Their navigation systems remain mysterious, although they are known to possess exceptionally good night vision and may be able to recognise their pond by smell.

Studies in which toads were marked and recaptured, including a long-term research project at Llandrindod Lake in Wales, have shown that many common toads return to the site where they were spawned. Not all toads do so, however. Some switch to a different pond, perhaps because they encountered it by chance while on their travels or because their original pond was in poor condition. Whichever the case, the result is that toads can colonize new breeding sites, thereby ensuring a population's survival.

FRENZIED MATING

Within a few days of the toads' arrival at a pond, the night air reverberates with the ventriloquial courtship croaks of the males. There are feverish scenes as rival males wrestle furiously to secure a mate, the aim being to grip their female partner tightly in a powerful embrace known as amplexus. By the end of April the mating season is over and the pond empty of toads, apart from the gelatinous strands of toadspawn attached to submerged plants.

After spawning, the toads move back to their feeding grounds to top up their severely depleted energy levels. They follow a less direct route than on their inward migration, but appear no less determined. In the Austrian Alps, scientists used radio-tracking to trace the post-breeding migrations of toads and discovered that some crawled up cliffs with 65° slopes in their effort to reach the best summer feeding areas. A few individuals climbed 400 m (450 yards), the highest vertical migration known for any amphibian. Meanwhile, back at the pond, the tadpoles take two to three months to develop into toadlets. The tiny toads clamber onto land and scatter into the surrounding vegetation. It will be at least two years (longer in females) before they make their first return migration to breed.

MIGRATION PROFILE

Scientific name	*Gecarcoidea natalis*
Migration	Between inland rainforest and the coast
Journey length	0.5–4 km (0.3–2.5 miles) each way
Where to watch	Christmas Island, Indian Ocean
When to go	November–January

Red Crab

On Christmas Island the first flush of the wet season triggers the migration of millions of red land crabs, which swarm from their rainforest homes down to the ocean to spawn. This living carpet of scuttling scarlet bodies surges back and forth several times over three lunar cycles.

Christmas Island lies to the south of Indonesia, a speck of green in the Indian Ocean. Its lush interior is cloaked with rainforest and teems with wildlife, but only three species of ground-living mammal are native to the island, two of which are extinct. Their place has been taken by no fewer than 13 species of land crab, and these crustaceans play a major part in the island's forest ecosystems.

KINGDOM OF THE CRAB

Red crabs dominate the terrestrial fauna of Christmas Island. They act as gardeners of the forest, aerating the soil by digging burrows, fertilizing it with their droppings, recycling nutrients by eating fallen leaves and fruit, and controlling the spread of undergrowth by feeding on seeds and seedlings. In the 1980s there were an estimated 130 million red crabs on the island, although this huge population shrank by 25 per cent during the mid-1990s due to a plague of introduced yellow crazy ants.

A few of the world's land crabs have evolved to breed in fresh water, but the majority, including the red crab, must return to the sea to spawn. The starting pistol for the red crabs' migration is the Indian Ocean monsoon, which usually hits Christmas Island in early November. The annual deluge immediately triggers increased activity in the crabs, whose behaviour is strongly linked to moisture – they are most active when the humidity soars above 70 per cent. The crabs leave their small home ranges and begin moving down from the forest plateau towards the coast, taking a week or so to arrive.

Researchers have followed the progress of individual crabs by means of colour-marking and radio-tracking. The studies revealed that the crabs travel in straight lines when possible, mainly during the cooler early morning. The maximum recorded distance walked by a red crab in one day was

Left The onset of the monsoon sends hordes of red crabs swarming across Christmas Island to their coastal breeding areas. They take the quickest route, undeterred by obstacles such as the island's railway.

AUSTRALIA

WAVES OF CRABS

The onset of the wet season in early November triggers the crabs' migration to the shore. They arrive in two main waves and dip in the sea to replenish essential mineral salts and water lost during the march. The males then move onto flat coastal terraces a short distance inland where they dig burrows, jousting with each other to occupy the best spots on the crowded terraces. As more females arrive in the second wave, mating takes place in or next to the burrows. The male crabs head back to the forest not long afterwards, leaving the females inside the burrows to brood their fertilized eggs.

The female crabs remain underground for about two weeks as their egg masses ripen, then emerge under the cover of darkness to cast the eggs into the sea, just as the high tide is turning during the last quarter of the Moon. Gathered on top of the low cliffs that encircle the island, the female hordes vibrate their bodies to flick their eggs over the edge. The eggs split open as soon as they tumble into the water, coating the surface of the sea with a blood-red slick made up of millions of microscopic crab larvae. Now the females can return to the forest to feed and recover.

New Moon
Full Moon
Last quarter of Moon
Phases of the Moon over a 3-month period from November to January
Females return to forest plateau
Young crab larvae mature in the sea, crawling on to land after about four weeks
Females and males
Males
Females
Crabs in first wave dip in sea before digging burrows
Second, larger wave of mostly female crabs
After mating, males head back to the forest
Females emerge from burrows to cast eggs into sea as high tide turns

1,460 m (1,600 yards), with the typical daily journey being less than half that; the longest migrations were just over 4 km (2.5 miles). It was suggested that the crabs navigate using a combination of visual clues, an internal magnetic map, and an ability to detect patterns of polarized light in the sky.

TIMED BY THE MOON

The spawning date of red crabs is governed by the lunar cycle (see box, above), with egg release taking place on four or five nights during the last quarter of the Moon. On these nights the height difference between high and low tide is smallest, which is critical because the female crabs can approach the shore safely, without getting washed away by a tidal surge.

Scientists comparing the red crab breeding seasons in 1993 and 1995 found that there was a three-week difference between the start of the migration in each year, but the spawning date remained fixed by the lunar cycle. In 1993 the crabs set off late, held up by the overdue arrival of the monsoon, so had to rush to reach the coast in time. In 1995, the crabs made a more leisurely journey, stopping off en route for up to a week to feed.

Many ocean-going plankton feeders time their December arrival off the coast of Christmas Island to coincide with the last quarter of the Moon, to plunder the dense clouds of crab larvae drifting with the current. The bonanza of food attracts whale sharks (see pages 94–95), manta rays, and large schools of fish. In some years very few baby crabs survive to return to shore, but an occasional good breeding season compensates for these losses.

Above Before mating, red crabs gather in sheltered bays, where they dip into the seawater to replenish essential salts. Later, each female will produce up to 100,000 eggs.

FROM EGG TO ADULT

A female red crab's eggs develop in brood pouches under her shell. The incubation period is 12–14 days. Once released into the sea, each tiny crab hatchling turns into a shrimp-like larva called a zoea. The zoea passes through several stages, moulting its rigid outer skin between each stage to match its growth, until three to four weeks later it reaches the final larval stage, called the megalops. The larva metamorphoses into an adult crab in the open sea. Although its shell is a mere 5 mm (0.2 inches) across, the small crab is strong enough to complete a nine-day migration to the rainforest.

Above Powered by rhythmic strokes of their massive tails, migrating humpback whales are able to cover vast distances with ease. In cruising mode, they travel about 120 km (75 miles) per day.

Migration in Water

Oceans and rivers support a staggering diversity of migratory creatures, including representatives from almost every major animal group. The smallest are tiny shrimp-like crustaceans barely visible to the naked eye, while at the other extreme, the colossal blue whale is the largest animal ever to have lived. Many aquatic migrants possess exceptional endurance and navigational abilities – turtles are able to pinpoint the precise part of a beach where they hatched years earlier, and sharks can home in on a submarine mountain or reef from thousands of miles away.

Humpback Whale

Humpback whales carry out the longest known migration of any mammal, from tropical waters to distant polar seas brimming with food, then back again. These creatures are renowned for their dramatic leaping displays and playful cavorting, and also for their haunting, extraordinarily complex courtship songs, which are performed by adult males.

MIGRATION PROFILE

Scientific name	*Megaptera novaeangliae*
Migration	From polar feeding areas to subtropical/tropical breeding areas
Journey length	Up to 8,500 km (5,250 miles) each way
Where to watch	Silver Bank, Dominican Republic; Byron Bay, New South Wales, Australia; Cape Cod, Massachusetts, USA
When to go	January–February (Silver Bank); Late May–July (Byron Bay); July–September (Cape Cod)

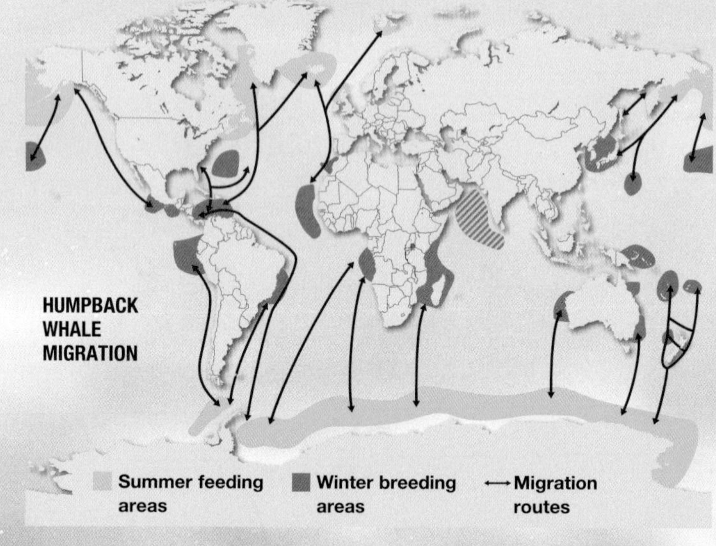

Acrobats of the sea, humpbacks seem to our eyes to be thoroughly enjoying themselves as they repeatedly smack the water surface with their flippers, roll onto their backs, or bring their tail flukes down hard to create a huge splash – a behaviour known as lobtailing. Without doubt their most spectacular move is the full breach, when a whale flings itself into the air and lands on its back. But in truth, these manoeuvres – which make humpbacks so exciting to watch – may have more importance as a means of socializing, foraging, and communication. One of their chief functions is to send messages about social status over long distances.

GREAT WHALES

Humpbacks reach 12–15 m (40–50 feet) in length, one-third of which is taken up by the enormous knobbly head, often liberally decorated with barnacles. They belong to the suborder Mysticeti, the baleen or so-called "great" whales, which contains the mighty blue whale and several other ocean giants. In common with the rest of this group, humpbacks are built for a life of long-haul cruising and on the upper jaw have a bristly fringe of baleen used to strain food from the water. Their throat has up to 36 deep pleats that stretch like a concertina during filter-feeding; over 2,000 l (500 gallons) of seawater and food cascade into the cavernous mouth with a single gulp.

For humpbacks, cold polar waters offer a feast worth travelling for. Here, nutrient-rich upwellings trigger an explosion of microscopic plant life that in turn sustains an amazingly productive ecosystem, from zooplankton to fish, squid, and larger predators. The whales' diet varies according to locality. Those in the Northern Hemisphere specialize in hunting small schooling fish, such as sand lance, capelin, mackerel, and herring. Southern Hemisphere humpbacks prefer to vacuum abundant swarms of zooplankton, especially shrimp-like krill. Humpbacks have developed a cooperative feeding strategy unique among whales: bubble-netting. Three or four will team up to herd fish or krill into a dense pack, then exhale rings of bubbles to surround the panicked shoal and block its escape. Groups of as many as 12 whales have been seen bubble-net fishing, but this is exceptional.

JOURNEY TO THE TROPICS

The feast is short-lived, as falling sea temperatures drive zooplankton to the seabed, where they become dormant. It lasts from around June to September in the Arctic Ocean and northern Pacific and Atlantic, and from around December to March in the Southern Ocean. Having eaten their fill and put on perhaps 10 tons of blubber, the bloated humpbacks make a beeline for warmer latitudes to mate and give birth. The population from each hemisphere visits the tropics and subtropics at a different time, so the two stocks seldom meet and are genetically quite distinct (although still the same species). The longest known humpback migration is between Antarctic waters and the sheltered Caribbean coast of Costa Rica.

One of the world's main humpback calving and mating grounds is Silver Bank, a sheltered stretch of shallow water north of the Dominican Republic in the eastern Caribbean. Up to 7,000 whales pass through this area during a three-month period. At least half of all humpbacks in the North Atlantic are thought to be born at Silver Bank, a marine reserve since 1986. Other important breeding areas include Hawaii, the west coast of Central America, the coast of west-central Africa near Gabon, and Australia's east coast.

SLOW RECOVERY

The first people to study the seasonal movements of humpbacks were 19th-century whalers, who soon knew exactly where to ambush them. Migrating humpbacks stick to predictable paths each year and tend to follow coastlines, which made catching them straightforward. More than 100,000 humpbacks were killed in the Southern Hemisphere in 1900–1940 alone. Today's steadily recovering global population stands at 30,000 and is probably a mere fifth of its original pre-whaling level.

Remarkably for such long-range migrants, humpbacks are often faithful to particular feeding and calving areas from season to season. Scientists know this because individual whales are easy to identify by the distinctive white markings on their flukes, flippers, and underside. There is now a database of photographs of several thousand humpbacks, including many from the North Atlantic population.

STRANDING

Thousands of cetaceans (whales, dolphins, and porpoises) are washed up on shores around the world every year. Among them are a few dozen humpbacks. Unable to refloat, most beached victims die of exposure, providing a windfall for carrion-eating animals. Stranding is a natural occurrence, typically involving old, sick, or injured individuals too weak to keep swimming, or inexperienced juveniles that have misread the changing contours of the seafloor or have strayed from their species' usual migration routes. However, reports of strandings are on the increase. This may suggest a change in the marine environment – caused by global warming or artificial underwater noise, for example – or it could be due to increased observer coverage.

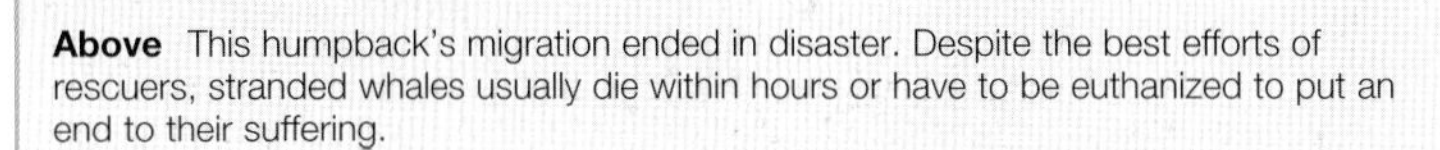

Above This humpback's migration ended in disaster. Despite the best efforts of rescuers, stranded whales usually die within hours or have to be euthanized to put an end to their suffering.

Facing page Humpbacks are named for habitually arching their back as they slip below the waves, but it is the attention-grabbing "full breach" that has made the species famous. **Facing page inset** Amid crashing spray and foam, several humpbacks round up a shoal of terrified fish on their summer feeding grounds.

MIGRATION PROFILE	
Scientific name	*Eubalaena australis*
Migration	From Antarctic feeding areas to temperate breeding areas
Journey length	Up to 2,500 km (1,500 miles) each way
Where to watch	Peninsula Valdés, Patagonia, Argentina; Hermanus Bay, Western Cape, South Africa
When to go	July–October

Southern Right Whale

As the plankton blooms of the Antarctic summer begin to dwindle, southern right whales turn north to warmer waters to mate and give birth. Their migration is fuelled entirely by fat laid down during their four-month sojourn in the icy Southern Ocean, since they will not feed until their return a year later.

Southern right whales are one of the world's few great whales to be increasing in number. Their total population was estimated at 3,000–4,000 in 2004, and continues to grow by 5 per cent a year. However, this is still a fraction of those alive prior to commercial whaling. The species' common name refers to the fact that these were considered the "right" whales to hunt because they were slow-moving and easy to chase, and had large amounts of blubber, ensuring they floated when harpooned. Many tens of thousands were slaughtered throughout the Southern Hemisphere between the late 1700s and the international whaling ban of 1935.

The sluggish movements of southern right whales, typically close to the surface, can be a good way to tell them apart from other baleen whales. Another clue to their identity is their habit of slapping the water with their tail and flippers, perhaps to help dislodge parasites or send signals to others. Their distinctive physical characteristics include deeply notched tail flukes and patches of paler skin on the head. Known as callosities, these crusty, barnacle-infested growths are unique to each whale, enabling researchers to recognize individuals and gradually to build up a picture of their daily and seasonal migration patterns.

COLD-WATER FEAST

When the whales arrive on their feeding grounds around Antarctica in December, at the start of the austral summer, they are often near starving. But the bountiful polar seas mean they can almost double their body weight in a short stay. The whales have an arched jawbone that serves as a huge feeding scoop. By cruising with it held partly open, they are able to skim the surface waters for zooplankton,

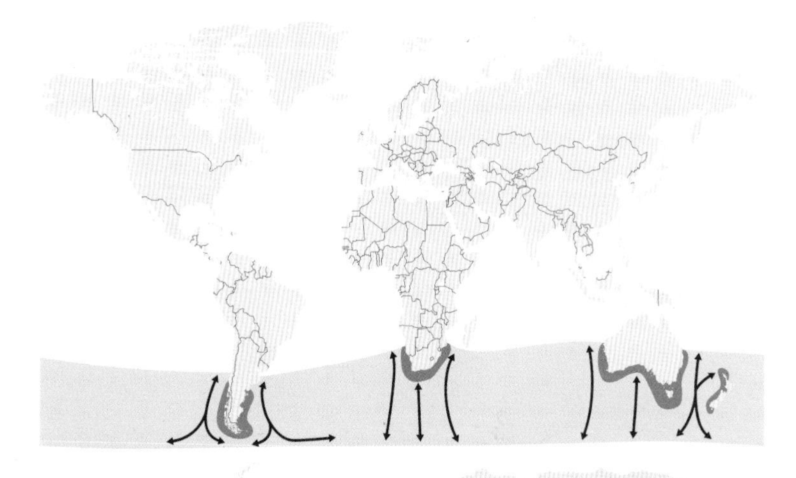

SOUTHERN RIGHT WHALE MIGRATION

Migratory range

Winter breeding areas

Migration routes

Below Mother and calf are inseparable for several years. Studies show that the female encourages play, which helps the calf to develop strength and swimming skills.

ATTENTIVE MOTHERS

Breeding is so demanding that a female southern right whale waits up to five years between calvings to recoup her energy. After a year-long gestation, the cow finally gives birth to a single calf about 5 m (16 feet) long and 1 ton in weight. Her rich milk is 40 per cent fat, yet despite being in the middle of her annual fast, she is able to produce enough for the baby whale to gain an average 60 kg (130 pounds) every day. The cow never leaves her calf's side for the 10–12 months it suckles, and the pair stick together for several more years. The strong mother-and-calf bond was exploited by whaling captains, who waited on sheltered coasts for heavily pregnant females to arrive and drop their babies. The cows would not desert their newborn young, making them sitting targets.

Facing page A southern right whale lifts its powerful tail flukes, revealing the distinctive notch in the middle that helps to identify the species.

especially krill and copepods (tiny swarming crustaceans). Curtain-like sheets of baleen suspended from the roof of their massive mouth trap the morsels of food for swallowing.

In March and April, the satiated whales return northwards. They retrace traditional routes – probably used by generations of whales – which hug coastlines to minimize the time spent in mid-ocean. As a result, the species is seldom encountered far from shore; few reliable records have come from the vast deep-water region between New Zealand and the Pacific seaboard of South America, for example. Normally the whales travel alone, in mother and calf pairs, or in small groups of three or four.

WARM-WATER NURSERIES

There are three main calving areas, which are located along the coasts of: Chile and Argentina; southern Africa; and Australia and New Zealand. These relatively warm waters are food deserts for the whales, but ideal for calving. Females seek out calm bays or lagoons to bear their young. An added advantage of the shallows is that marauding great white sharks and pods of orcas pose much less of a threat to the vulnerable baby whales. Favoured locations, such as Peninsula Valdés in Patagonia, or the series of bays to the east of the Cape Peninsula in South Africa, host loose congregations of dozens or even hundreds of whales, and boast unrivalled opportunities for shore-based whale watching. The whales are present from May to about November, and include females with calves in tow, clusters of rival males competing for a mate, non-breeding females, and juveniles (those under 10 years of age).

Mating takes place in July–October, depending on the locality. A curious aspect of right whale anatomy is that the brain of an adult male weighs approximately 4 kg (9 pounds), while his testes weigh as much as 1,000 kg (2,200 pounds). They are the largest gonads in the animal kingdom. Maintaining the sexual organs is energetically expensive, so like other baleen whales, the male has shrunk his brain to compensate.

ENDANGERED RELATIVE

The Northern Hemisphere is home to two closely related species of whale, virtually identical to their southern counterparts in looks and behaviour. As might be expected, the migratory pattern in northern right whales, *Eubalaena glacialis* and *E. japonica*, is reversed. Those in the northwest Atlantic (*glacialis*) commute between food-rich waters near eastern Canada and New England, and southerly calving grounds off the coasts of Georgia and Florida. The movements of whales in the northern Pacific (*japonica*) are barely known, although the Bering Sea is an important feeding area. Both populations of northern right whales are now seriously endangered. Fewer than 350 individuals survive in the Atlantic and only a handful are left in the Pacific.

Grey Whale

East Pacific grey whales undertake one of the planet's great marine migrations, between their northern feeding grounds and their winter calving and mating areas in the warm lagoons of Baja California, Mexico. Hugging the coast, these barnacled leviathans travel slowly but surely, arriving several months later.

Below Grey whales tolerate boats and divers at very close range, which has helped scientists to develop an intricately detailed picture of their migrations.

MIGRATION PROFILE

Scientific name	*Eschrichtius robustus*
Migration	From Arctic/subarctic feeding areas to temperate/subtropical breeding areas
Journey length	Up to 8,000 km (5,000 miles) each way
Where to watch	San Ignacio Lagoon, Baja California, Mexico
When to go	January–March

Grey whales are medium-sized, blunt-headed members of the suborder Mysticeti – the baleen whales. On average they are 13–14 m (43–46 feet) long, roughly the same as humpbacks, the species with which they are most likely to be confused. It is unusual for migrating grey whales to stray more than a few miles offshore. They are creatures of shallow seas, favouring waters under 60 m (200 feet) deep, and frequently pass close to cliffs and headlands on their marathon swim. One of their main feeding areas is the northern Bering Sea and the adjacent Chukchi Sea, where the continental shelf is both broad and extremely shallow (during the last ice age, when sea levels were lower, it formed a land bridge joining Asia and North America).

For a while in the Arctic summer, these cold, super-productive waters host about 2,000 grey whales. A further 20,000 or so whales will be scattered along North America's western seaboard, mainly off the coast of Alaska and British Columbia, although increasing numbers venture no further north than Oregon and California. This healthy population represents an astonishing comeback for a species that had been hunted to the verge of extinction by 1930, when only a few hundred individuals were thought to be left in the whole of the eastern Pacific.

However, the western Pacific stock of grey whales is critically endangered. Possibly as few as 100 survivors still make the passage from their summer home in the Sea of Okhotsk, which lies between Japan and Russia's Far East, to their traditional breeding areas in the waters of the Korean Peninsula. Oil and natural gas exploration continue to degrade their coastal habitat.

MUD GRUBBERS

Uniquely among baleen whales, greys harvest the seafloor. They swim low over the bottom like dredgers, scraping up huge mouthfuls of mud to sift out the hidden mass of tubeworms and small, shrimp-like crustaceans called amphipods. The seabed is left with tell-tale gouges similar to the furrows of a ploughed field. In fact, the whales' feeding action is indeed comparable to farmers tilling the land: by churning up the sediment, they help nutrients to cycle, which enriches the marine ecosystem.

Occasionally the whales use their stiff baleen plates to comb delicately through forests of kelp to strip encrusting crustaceans off the fronds. They also have several more orthodox feeding techniques. In common with other baleen whales, they cruise at the surface to sieve swarms of tiny crab larvae and krill from the water, and plunge into schools of small fish, jaws agape.

During an intense foraging season lasting up to five or six months, the whales lay down sufficient reserves of blubber to last the remainder of the year. This includes the significant energy cost of return migration. Not only do the whales fast for much of the journey, but also throughout their brief stay on the southern calving grounds from January to February or March, when there is little or no suitable food.

SLOW PROGRESS

Grey whales head south in October, November, or early December, depending on the latitude of their summer range; those in the Arctic set off first. The trigger to depart could be the fall in sea temperature, build up of sea ice, and shortening daylight hours. The slowest of all whales, greys move at no more than 7–9.5 kph (4–6 mph), and females with calves are somewhat slower. In addition, the whales take regular rests. Whale lice and barnacles therefore spread freely across their body, which adds to the species' characteristically mottled, scarred appearance.

Observations at sea and data from satellite-tagged whales suggest that the grey whale migration advances at 65–80 km (40–50 miles) a day. Until recently, greys were thought to migrate further than any other whale, but it now seems that few of this species complete the longest possible route from the Arctic to Mexico every year, and the total annual distance travelled by Antarctic humpbacks is greater. Since greys are strictly coastal, they probably navigate in part using visual clues above and below the waterline, such as landmarks on shore and changes in the undersea terrain. This could be the purpose of the manoeuvre known as the spy-hop, in which a whale hangs vertically in the water with its entire head lifted clear, in periscope fashion, and looks around.

KILLER WHALES

Lacking speed, migrating greys are vulnerable to attacks by pods of orcas. Their best means of escape are to hide in dense kelp forests or among rock formations near the seabed, neither of which is possible in deeper waters, such as submarine canyons. In the 1980s, scientists identified a genetically distinct population of about 400 orcas that hunt marine mammals in the northern Pacific. In April, they migrate to the Californian coast to intercept female and newborn greys heading northwards.

WHALE POUNDS

Conservationists often remark that today whales are much more valuable alive than they ever were dead. The first organized excursions to observe grey whales took place in California the early 1950s, spawning a global whale-watching industry now worth over £0.5 billion. Each year an estimated 11 million people go on whale-watching trips in 87 countries worldwide.

Right These are just a few of the 100,000 eco-tourists who visit the grey whale calving lagoons in Baja California every year.

Walrus

Each summer walruses migrate north as the Arctic pack ice withdraws, travelling in single-sex herds to dive for mussels and clams in the productive polar waters. When the ice sheet advances again in autumn, the corpulent mammals beat a retreat to their mating and wintering areas further south.

Walruses are heavyweights of the suborder Pinnipedia, or seals and sealions, being second only to elephant seals in size. Adults weigh in the range 1,250–1,700 kg (2,750–3,750 pounds), with the bulls 50 per cent heavier than the cows, and their thick-skinned, wrinkly, grossly distended bodies are quite unlike the smooth, tapering silhouette of other pinnipeds. Appropriately, the name walrus is said to be derived from the old Dutch for "shore giant". Despite their bulk and clumsiness on land, walruses are powerful swimmers, capable of speeds of at least 35 kph (22 mph). They can be surprisingly graceful underwater, and groups of juveniles engage in playful chases and frolics in the shallows.

The most obvious feature of walruses is their tusks – actually modified canine teeth. Present in both sexes, these dagger-shaped appendages were long believed to be used for dredging bivalve molluscs – the favourite food of walruses – from seabed mud, but in fact serve as status symbols, used in threat displays. Older, high-ranking individuals own the biggest tusks, while immatures have none. Tusks also help walruses to grip slippery ice floes, and deter polar bears (see pages 46–47), although bears seldom chance an attack against such formidable opponents unless desperate for a meal.

FORAGING CYCLE

Walruses have a circumpolar distribution in coastal waters around the Arctic, with three discrete populations in the Pacific, Atlantic, and Siberia's Laptev Sea. Their migration patterns vary, but most appear to make two types of journey: short-term and long-term. For much of the year their routine involves short trips over the continental shelf to find food, lasting up to four or five days, followed by one or two days spent loafing on sea ice or beaches. This foraging cycle is repeated over and over, as walruses work to harvest the glut of shellfish. Researchers found that during a typical dive of around 5–7 minutes, a walrus consumes more than 50 mussels, equating to just over 73 kg (160 pounds) of mussel flesh per day. Walruses use their rest days to digest and socialize.

Above Each year, walruses travel to ancestral haul-outs to moult. The dense crowds have an important social function, as walruses gain reassurance from close physical contact.

The longer, seasonal walrus migrations are dictated by the ebb and flow of sea ice. Even though walruses can dive beneath ice to plough the seafloor for buried shellfish, and are able to smash breathing holes in it, they prefer areas with scattered floes and avoid those that are heavily ice-bound. The best-studied migratory population is the Pacific subspecies. Having wintered in the Bering Sea, these walruses start to move north through the Bering Strait in about April. They travel in separate male and female herds up to 50 strong, and the females give birth while on the migration, usually in May–June. The walruses remain on their summer feeding grounds in the Chukchi Sea, along the edge of the permanent ice cap between Russia's Far East and the northwest corner of Alaska, from July until September, before returning south via the Bering Strait. Many of the bulls summer further south, and meet up with the cows and their calves in winter. It is thought that sexual segregation enables walruses to reduce competition for food during the crucial summer fattening period.

LIFE'S A BEACH

Walruses, in common with most seals and sealions, have not entirely lost their land-living ancestry, because they are unable

Above The ebb and flow of sea-ice is the reason walruses evolved a migratory lifestyle. However, their ice-dependence now leaves them vulnerable to global warming.

to give birth at sea and must come ashore to mate and moult their dead skin. This forces them to travel to traditional haul-outs at certain times of year. Suitable sites have a gently shelving sea floor, a beach protected from crashing surf, and cliffs on the landward side as defence against land predators. The largest haul-out in the Western Hemisphere is Round Island in Bristol Bay, Alaska, where in early summer 2,000–10,000 walruses crowd together like a horde of bloated sunbathers. They flush bright pink in the warm sunshine as blood rushes into their thick blubber to help them cool down.

MIGRATION PROFILE

Scientific name	*Odobenus rosmarus*
Migration	To northern feeding grounds in summer
Journey length	Up to 3,200 km (2,000 miles) annually
Where to watch	Round Island, Alaska, USA; Spitsbergen, Arctic Ocean
When to go	June–July

WARMING THREAT

In the past, walruses were slaughtered in huge numbers for their blubber, ivory tusks, and abundant body fat, which was rendered into oil. They are still taken by indigenous Arctic peoples, such as the Chukchi of northeast Siberia and the Yupik of western Alaska, but climate change has replaced hunting as the main threat to the species, with forecasts that rising sea temperatures will cause a rapid shrinkage of polar ice cover. Females with calves haul out on ice rather than the shoreline, so loss of sea ice is affecting survival of the young. Moreover, the zone of broken sea ice that forms the summer habitat of walruses is receding slowly northwards away from the coastal shallows into deeper waters. This could put the sea bottom beyond their reach, since walruses dive most efficiently in water 10–50 m (30–160 feet) deep.

Above Walruses have a good sense of smell but poor eyesight, so can easily be approached from downwind.

Magellanic Penguin

Named after the Portuguese explorer Ferdinand Magellan, this penguin lives in the cold waters around the remote southern tip of South America. Until recently its movements at sea were a mystery, but new research suggests it undertakes some of the longest journeys of any flightless bird.

The Magellanic penguin is one of four penguins to nest on the South American mainland, and this species can be distinguished from its relatives by the two black bands across its chest. It is an abundant penguin, with a total population estimated at 1.8 million pairs, split between Chile (800,000 pairs), Argentina (900,000 pairs), and the Falkland Islands or Islas Malvinas (100,000 pairs), which lie in the South Atlantic to the east of the Argentine coast. In the north of its range the species is mostly sedentary, remaining in the vicinity of its breeding colonies all year, but further south it is a great wanderer.

Each year birds from the Magellanic penguin's southern colonies swim hundreds of miles between leaving their breeding grounds in March or April and returning again some time in August or September. Imagine riding the mountainous swell of a storm-buffeted ocean, far from shore in the darkness of a long winter night, and surviving this challenge not just once but many times, all during the Southern Hemisphere winter. On top of this, the penguins navigate across the featureless surface of the sea without the benefit of being able to scan the horizon for landmarks, like migratory birds flying high above the waves. It is truly a formidable achievement.

PRODUCTIVE SEAS

The largest colony of Magellanic penguins – indeed, of any penguin outside Antarctica – is at Punta Tombo in Argentina. This barren spit of rock and shingle provides a stark contrast to the tremendously productive ocean around it. A vast continental shelf extends into the Atlantic for 480 km (300 miles), as far as the Falklands, and its shallow, nutrient-rich waters teem with fish and squid. The area supports amazing concentrations of wildlife in summer. Besides penguins, the beaches at Punta Tombo are used as haul-outs by southern sealions, Antarctic fur seals, and southern elephant seals. Offshore there are southern right whales (see pages 80–81) and pods of orcas.

Magellanic penguins are creatures of habit: they form lifelong pair-bonds and most are faithful to the same nest site within their colony. Both sexes share the task of incubating their two eggs, for a total of about 40 days, and go fishing daily for the fast-growing chicks. The penguins target shoals of small fish such as sardines and often are joined by seals, sealions, and albatrosses, so the panicked fish end up suffering a multipronged assault from both water and air, making them easier to catch.

WINTER DISPERSAL

After the breeding season, the penguins head out to sea. Those from Punta Tombo follow the coastline northeast, towards the waters off Uruguay and southern Brazil where they spend the winter. They swim with the cold Falkland Current that surges northwards to the estuary of the Río de la Plata, at which point it meets the warmer, southward-flowing Brazil Current. Here, the mixing of cold and warm water provides plenty of food.

Penguins from the Falkland Islands population first move northwest towards the coast of Argentina, then follow its coastline like the Punta Tombo birds. Magellanic penguins that breed on the Chilean coast also head north after breeding: they swim with the mighty Humboldt Current, which flows up the west coast of South America towards southern Ecuador.

YOUTHFUL STRAYS

There are important differences between the dispersal of adult and juvenile Magellanic penguins. Whereas adults stay in the breeding area for a few weeks to moult their plumage, the juveniles leave almost immediately and moult at their winter feeding grounds, gathering in dense flocks along the shore. Young penguins appear to have a powerful urge to wander: there are numerous sightings of them huge distances from the nearest breeding colony. Juvenile Magellanic penguins have been spotted on the other side of the world, off the coast of Australia and New Zealand.

MIGRATION PROFILE

Scientific name	*Spheniscus magellanicus*
Migration	Disperses north after breeding
Journey length	100–1,000 km (60–600 miles)
Where to watch	Punta Tombo, Argentina; Falkland Islands, South Atlantic
When to go	November–January

MAGELLANIC PENGUIN MIGRATION

→ Winter dispersal
Total range
← Summer migration to breeding grounds

TRACKING PENGUINS

In March 1998 scientists on the Falkland Islands fitted lightweight devices called platform transmitter terminals (PTTs) to 10 Magellanic penguins to track their post-breeding movements. In this particular sample, the longest distance travelled by an individual Magellanic penguin was 2,661 km (1,653 miles) within 75 days. Today, more sophisticated electronic tags provide data such as dive depth, energy expenditure, and feeding behaviour.

Above Our understanding of penguin dispersal at sea has been transformed by satellite transmitters.

Below Argentina's wild southeastern coast is a breeding ground for Magellanic penguins. They are attracted by the shallow waters that lie offshore, which form one of the world's most fertile fishing grounds.

Atlantic Ridley Turtle

Virtually the entire world population of Atlantic ridley turtles nests on a single Mexican beach on the same day. The females haul ashore to lay en masse, after navigating to the precise spot from their feeding grounds in the Gulf of Mexico and within 12 hours all have gone.

Atlantic ridleys, also known as Kemp ridley turtles after Richard Kemp, the Florida fisherman who first presented the species for scientific examination in 1880, have an almost circular, greyish-green carapace and a distinctive parrot-like beak. They are the most endangered of the world's seven marine turtles, or cheloniids, and the smallest, measuring only 66–70 cm (26–28 inches) long.

Like their relatives, Atlantic ridleys are strongly migratory, and the adult females commute between feeding grounds and traditional nesting beaches. However, they differ from the rest of their family in several important respects. Their breeding area is extremely localized, confined to a short strip of coastline in the western Gulf of Mexico, whereas most other sea turtles have global ranges and nest in oceans around the tropics. Females usually lay eggs in broad daylight, not under the cover of darkness. And many of the adult males appear to be non-migratory, having lost their urge to wander.

NESTING MULTITUDE

Atlantic ridleys provide arguably the ultimate example of synchronized reproduction in vertebrates. Dozens of egg-bearing females emerge from the sea simultaneously during a mass nesting called an arribada, from the Spanish for "arrival". They clamber over each others' shells, flippers flailing, sometimes in the confusion digging up the eggs of females that went before. Olive ridleys are the only other sea turtles to perform arribadas, but they have a

MIGRATION PROFILE

Scientific name	*Lepidochelys kempii*
Migration	To and from nesting beaches
Journey length	Up to several thousand miles a year (adult females)
Where to watch	Rancho Nuevo, Tamaulipas, Mexico; Padre Island National Seashore, Texas, USA
When to go	May–July (restricted access)

Facing page With its front flippers beating like wings, a turtle seems to fly through the water. The downstroke generates most of the forward thrust. **Right** Illuminated momentarily by the photographer's flashgun, the latest crop of hatchlings scrambles down the beach to the sea.

pantropical distribution, so their breeding is dispersed over a wide area. In contrast, over 95 per cent of Atlantic ridleys converge on a small part of Tamaulipas state in northeast Mexico, especially a remote 20-km (12-mile) beach at Rancho Nuevo. The remaining 5 per cent nest in neighbouring Veracruz or in Texas, with a handful (often fewer than 10 turtles) heading to Florida.

A famous film of Atlantic ridleys pouring onto the beach at Rancho Nuevo, shot in 1947, has enabled scientists to calculate that 40,000 females made it onto the beach that day. The spectacle was then one of the planet's greatest wildlife phenomena. It is much diminished today, due to the species' decades-long population decline. The female turtles come ashore two or three times between May and July, and during each arribada lay a clutch of 90–100 eggs in deep pits, disappearing into the water again by early evening. The eggs require 42–60 days' incubation, depending on the temperature. Over several nights the thousands of 3.8-cm- (1.5-inch-) long hatchlings race to the water's edge to begin a nomadic life in the open sea.

JUVENILE DISPERSAL

The hatchlings swim strongly to reach the relative safety of deeper water, where they drift with ocean currents for probably the next two or three years. It seems likely that, in common with juveniles of other sea turtle species, they frequent mats of floating sargassum seaweed, which would offer plentiful small morsels of food and serve as a refuge from predatory fish. Some of the young turtles might remain within the Gulf of Mexico, while others are swept out into the western Atlantic by the powerful Gulf Stream. Upon reaching sub-adult size, they return to shallow nearshore waters, such as the mouths of the Mississippi and Alabama rivers. Here, they forage for crabs, shrimps, and bivalve molluscs on sandy or muddy seafloors, or among eelgrass beds.

Male turtles will spend the rest of their lives over the continental shelf: some move up and down Gulf coasts and the USA's Atlantic seaboard, north as far as New England, in search of the best feeding; others, possibly the majority, stay faithful to the same area. It is not known where mating occurs, although the females typically embark on their first nesting migration when about 12 years old. As few as one in a thousand are estimated to survive to adulthood and make it back to their natal beach to lay.

For female Atlantic ridleys to lay in tandem, there must be reliable environmental or physiological triggers. Cues for their arribadas could include the lunar cycle and tides, seasonal changes in offshore winds, or the release of pheromones by females. Not every mature female takes part, however. Most rest for one or two years before mating again. It is thought that the purpose of arribadas is to minimize the impact of egg thieves, such as crabs, seabirds, and vultures, by suddenly overwhelming them with potential food.

BACK FROM THE BRINK

Collecting Atlantic ridley eggs was banned in 1966, but nevertheless by the late 1970s, the total number of nests had fallen to an all-time low of only 700 per season, produced by several hundred surviving adult females. Since 1978, a partnership of Mexican and US government agencies and NGOs has maintained an intense conservation effort to lift the species from near-extinction. Regular patrols of the main beach at Rancho Nuevo mean that most clutches can be translocated to artificial nests. Thousands of eggs have also been taken to Padre Island, in Texas, to sow the seeds for a second major breeding population. The eggs are raised in incubators, using a temperature that ensures most hatchlings are female.

Left At Rancho Nuevo, hundreds of artificial nests are cordoned off to protect the developing turtle eggs from predators and poaching.

Green Turtle

For centuries or even millennia, green turtles have travelled to the same beaches to lay their eggs. Driven by an amazingly accurate homing instinct, they can navigate to tiny islands in mid-ocean, finding the very part of the hatchery where they started life many years earlier.

Green turtles are named after the colour of their body fat, but in fact their most distinctive feature is the pattern formed by the pale edges to their scutes, or shell plates. They are among the largest sea turtles, usually 0.8–1 m (2.5–3.25 feet) in length, with a lightweight, beautifully streamlined carapace that greatly reduces drag in the water. It is the long front flippers that propel them forwards, in the manner of beating wings, while the squarish, webbed rear limbs serve as rudders. When fully grown at 40–50 years old, these turtles are strong enough to swim against currents if necessary, and can complete exhausting migrations that last weeks.

In much of the world, green turtles are the commonest of the seven ocean-going turtles in the order Chelonia. Although numbers have fallen due to commercial egg harvesting and hunting, the World Wildlife Fund (WWF) nevertheless estimated there to be 203,000 nesting females in 2005–2006, and the true figure may be much higher. In the case of such a wide-ranging migrant, found in tropical and warm-temperate seas worldwide, it might be more useful to list individual populations as endangered and not the entire species: green turtles underwent a rapid decline in the Caribbean, yet are thriving in the South Atlantic, where their numbers have increased by 300 per cent since the 1970s.

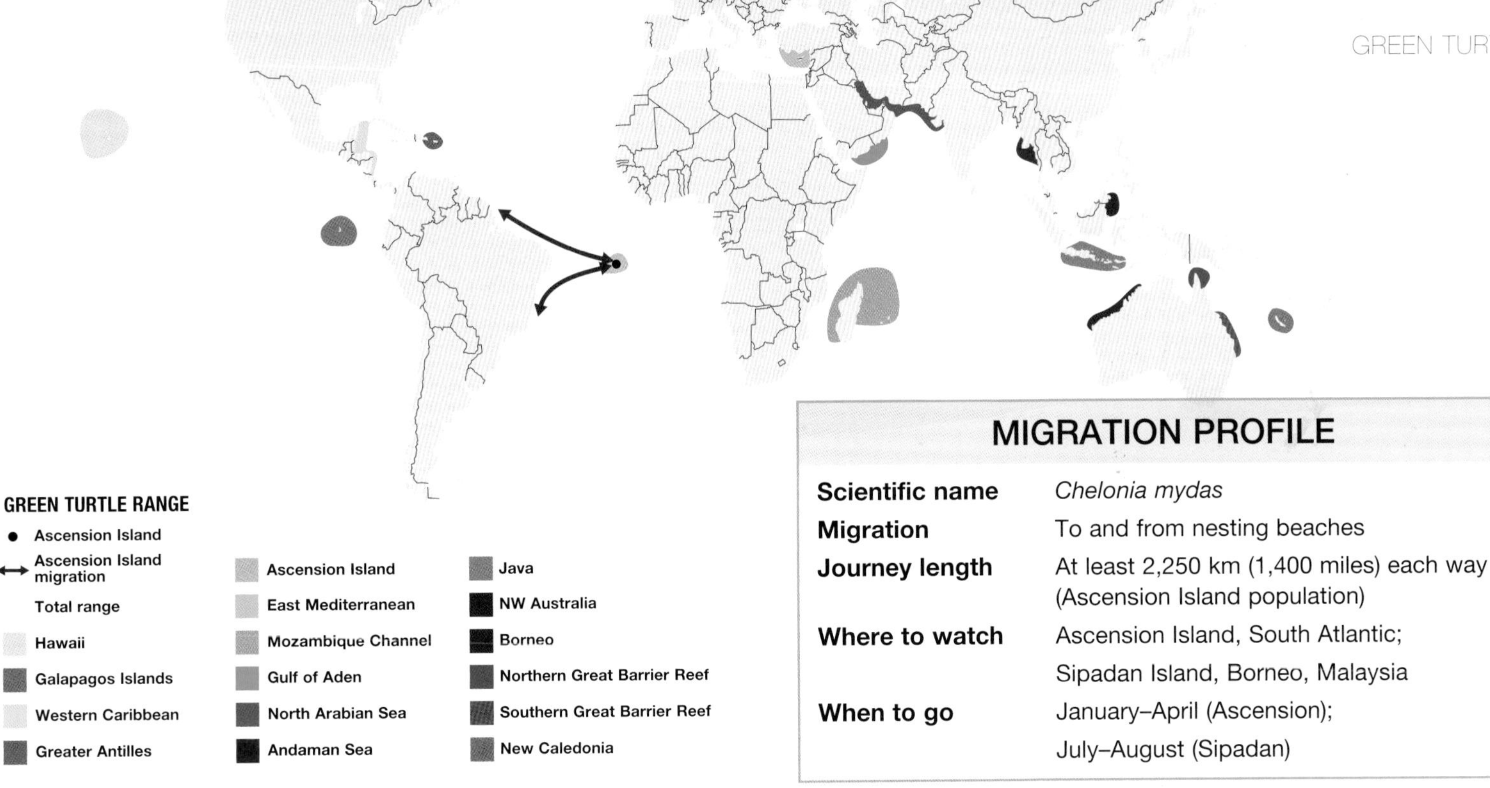

MIGRATION PROFILE

Scientific name	*Chelonia mydas*
Migration	To and from nesting beaches
Journey length	At least 2,250 km (1,400 miles) each way (Ascension Island population)
Where to watch	Ascension Island, South Atlantic; Sipadan Island, Borneo, Malaysia
When to go	January–April (Ascension); July–August (Sipadan)

VOLCANIC OUTPOST

Green turtles have evolved a strategy of nesting mainly on remote islands, where there are fewer predators to plunder their eggs and vulnerable, soft-bodied young. Their traditional nesting grounds are called rookeries. One of the best-known is Ascension Island – a mere pinprick only 13 km (8 miles) wide, located in the mid-Atlantic halfway between Brazil and the west coast of Africa. To get to this volcanic outpost, the turtles migrate from feeding areas in Brazilian coastal waters, a distance of 2,250 km (1,400 miles) or more. Because they graze almost exclusively on algae along coral reefs and on seagrass beds in sheltered bays, they have to go without eating en route. Moreover, their favourite plant food is scarce in the waters around Ascension, so their fast may last until they return to Brazil months later.

Both sexes make the trip to Ascension, and mating takes place in the shallows just off the nesting beaches. Each female is sexually receptive for about a week, during which she will copulate several times with different partners, storing their sperm. Soon afterwards, she comes ashore at night on the high tide, to help her cross reefs fringing the coast, and crawls to the top of the beach, safely beyond the strandline. Here, she lays 80–150 eggs the size of table tennis balls, burying them in the sand. The turtle may repeat this arduous operation up to nine times during the nesting season (January–April at Ascension), but three clutches is normal.

Female green turtles nest at intervals of two to five years. However, in common with other slow breeders, they are exceptionally long-lived animals: their average lifespan is thought to exceed 80 years.

Facing page Green turtles can be surprisingly approachable, but if spooked by divers they will put on a sudden burst of speed and disappear into the blue in seconds.

HOMING INSTINCT

Green turtle rookeries are often so small and isolated that the turtles' instinctive ability to pinpoint them across huge stretches of open ocean has long fascinated scientists. The turtles probably navigate by the position of the Sun and stars. They might also "read" wave and current patterns or follow the thermal gradients caused by local differences in water temperature. Another possibility is that they are able to sense variations in the Earth's magnetic field, forming a "magnetic map".

In one experiment, seven green turtles that had finished breeding on Ascension Island were fitted with electronic tracking devices, and with magnets to upset magnetic navigation. These magnetically disrupted turtles were then tracked by satellite as they returned to Brazil. They managed to keep to a course as efficiently as turtles not wearing magnets, proving that a geomagnetic sense, although it may play a part, is not essential to their migration.

DASH TO THE SEA

Green turtle eggs hatch after 45–70 days, according to the temperature inside each nest. Since turtles lack chromosomes that determine sex, the warmth of the sand also decides the sex of the developing embryos; males predominate below 29°C (84°F), and females above 31°C (88°F). By emerging together at night the hatchlings avoid seabirds and most other predators.

Above The hatchlings are guided by varying light levels on the horizon, which appears brighter over the ocean due to reflection from the water surface – even at night or in cloudy conditions.

Left Baby turtles crack their leathery eggs using a spike on their snout called an egg tooth.

Magnetic Attraction

At certain points across the globe ocean-going sharks materialize out of the blue in shoals of thousands, stay awhile, then disappear to destinations unknown. One such migration hotspot is Cocos Island, a solitary speck in the East Pacific.

Cocos is a remote seamount, or submarine mountain, located 550 km (340 miles) southwest of Costa Rica, to which it belongs. Regarded as one of the world's top dive sites, it has become a mecca for shark enthusiasts due to the regular occurrence of huge schools of scalloped hammerheads in the deep waters that surround its steeply shelving flanks. Similar assemblies of hammerheads take place at other seamounts in the East Pacific, such as El Bajo Espiritu Santo in the Gulf of California. Marine biologists think that the sharks, far from being evenly spread throughout the ocean, use seamounts as stepping stones on relatively quick migrations along fixed underwater highways. The stop-offs probably serve an important social function, enabling sharks to mingle and encounter potential mates.

During their visit to Cocos Island, the sharks make daily movements governed by the light–dark cycle. They spend the day resting offshore in packs, then at nightfall small groups disperse up to 16 km (10 miles) away to hunt shoals of squid, returning just before dawn. An experiment at El Bajo Espiritu Santo, in which hammerheads were tagged with ultrasonic transmitters, demonstrated that every morning individuals would home to within about 230 m (750 feet) of their place of departure. The sharks may be able to orientate by following "roads" of magnetism in the seafloor. In fact, entire seamounts could act as magnetic beacons, because they are made of volcanic rock and therefore have a pair of oppositely charged poles.

Left Scalloped hammerhead sharks rest and socialize in the waters around Cocos Island during a break in their oceanic migration.

Below left The underside of a shark's snout is covered with tiny pits, called ampullae of Lorenzini, which contain electromagnetic receptors used in orientation and hunting.

Below centre It is thought that the peculiar head shape of hammerhead sharks acts as a powerful scanner, enabling them to detect signals from prey hiding in sandy seabeds.

Below right This false-colour image, produced using echo soundings, shows a line of seamounts off Costa Rica. Seamounts are extinct volcanoes on the seabed.

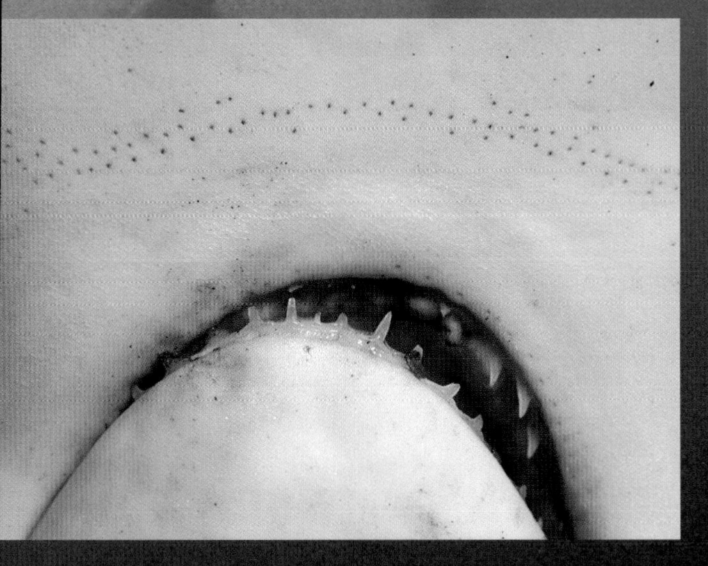

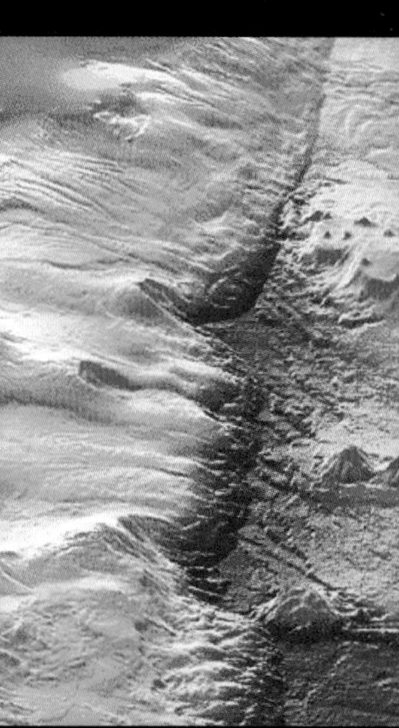

Above Individual whale sharks can be recognized from their tell-tale patterns of white spots, giving biologists new insights into the movements of these magnificent creatures.

Whale Shark

Giants of the deep, whale sharks are the world's largest fish. They are filter feeders that travel ceaselessly between parts of the ocean rich in plankton, faithfully turning up at these seasonal hotspots year after year to vacuum up great mouthfuls of the murky soup.

CAVIAR OF THE REEF

Whale sharks make routine trips to gorge themselves on coral eggs, the so-called "caviar of the reef". Simultaneous mass spawning events, triggered by the full moon, suddenly flood the sea with vast quantities of pink or red coral eggs. Water currents whip this into a thick, protein-rich slick that may stretch for hundreds of miles, since coral reproduction is synchronized along entire reefs.

Left Corals reproduce sexually, and their free-floating eggs are one of the favourite foods of whale sharks.

In contrast to the slim profile typical of many ocean-going sharks, a whale shark looks distinctly bulky. Instead of a conical snout, it has an enormously heavy, flattened head ending in a mouth up to 1.8 m (6 feet) across – comfortably wide enough for a person to swim inside. At around 12 tons and 12–20 m (40–65 feet) in length when fully grown, the shark would be unstable in the water were it not for three key adaptations. Pronounced ridges down the rear half of its body and tail stem, or peduncle, are stabilizers that prevent it from rolling from side to side as it cruises along, while the outsized tail and pectoral fins provide additional balance.

PLANKTON FEAST

Whale sharks usually feed passively, patrolling just beneath the surface with their mouth agape. They draw in huge loads of water, which flows across their five large gill slits, where food particles such as the eggs and larvae of invertebrates and fish are sieved out by a fine mesh of tooth-like bony projections ready to be swallowed. (Their true teeth are minute and play no part in feeding.) Only two other sharks – the basking and megamouth sharks – feed in this manner, and all three species are highly migratory. Megamouths are extremely rare and poorly known, but the life of basking and whale sharks can be characterized as an endless commute between the food "pulses" that develop in different locations at different seasons, and they may not get the chance to feed during their often lengthy journey to reach each area.

Above Whale sharks are mobile habitats in their own right, more often than not accompanied by colourful shoals of trevallies and other fish.

MIGRATION PROFILE

Scientific name	*Rhincodon typus*
Migration	Follows seasonal appearance of plankton blooms
Journey length	Probably several thousand km a year
Where to watch	Bay Islands, Honduras; Ningaloo Reef, Western Australia; Mahé Island, Seychelles; Great Barrier Reef, Queensland, Australia
When to go	February–April (Bay Islands); March–April (Ningaloo Reef); October (Mahé); November–December (Great Barrier Reef)

OCEANIC CURRENTS AND WHALE SHARK RANGE

Total range **Cold currents** **Warm currents**

Note: Other large, ocean-going fish, such as manta rays, make similar movements, as yet unknown.

Two active foraging techniques have been noted in whale sharks. Sometimes they hang vertically in the water to suck up shoals of small fish, tropical krill (shrimp-like crustaceans), or jellyfish from surface waters, and occasionally several will work together at the surface to corral their prey into dense packs. Normally, however, whale sharks lead a solitary existence.

HEAT SEEKERS

The global distribution of whale sharks is strongly influenced by warm ocean currents. They avoid waters cooler than 20°C (68°F), which ties them to tropical and subtropical seas, and prefer a surface temperature of 21–26°C (70–79°F). Most sightings occur in the lagoons of coral atolls and along coastal reefs, perhaps indicating that they hop from one favoured area to another and use the deep blue only when in transit, but this bias could simply reflect the difficulty of locating the sharks in the open ocean. Some of the best places to watch whale sharks include Ningaloo Reef in Western Australia, the Seychelles and Maldives in the Indian Ocean, and islands off the coasts of Honduras and Belize in the western Caribbean. The sharks tolerate divers and boats at close range, which has led to the growth of a shark-encounter industry worth an estimated £25 million annually worldwide.

It was first suggested in the 1930s that whale sharks undertake long migrations, breeding in the Indian Ocean, then heading south in the Mozambique Current around South Africa and out into the Atlantic, where they would ride the Southern Equatorial Current all the way to the Caribbean. This has never been proven, and yet satellite-tagged individuals have indeed travelled great distances; one shark in the Pacific covered 22,500 km (14,000 miles) in 40 months. But some scientists now argue that the sharks are more likely to clock this up over the course of a series of shorter migrations within an ocean basin, rather than with huge trans-ocean journeys. For example, large numbers of whale sharks gather off southeast New Guinea in September, before travelling south to the Great Barrier Reef in November–December, and then continuing southwards down the east coast of Australia (the return migration to New Guinea remains a mystery). Similar regional movements probably occur in the Gulf of Mexico, Indian Ocean, and Pacific.

Whale sharks are born with unique patterns of white blotches on their backs, which like human fingerprints do not change with age, and this allows researchers to recognize individual sharks so that their movements can be plotted. In the past, this analysis was carried out by laboriously studying photographs taken at sea. Today, NASA-designed star mapping software, originally developed for cataloguing galaxies, is used to identify sharks from their photographs. The research has shown that whale sharks display a high degree of philopatry – that is, they are very faithful to their main feeding locations, arriving within a few weeks of the same date each year, almost as though regulated by clockwork.

Blue Shark

With no home territory to speak of, blue sharks are long-range fish that patrol vast sweeps of the planet's seas, making seasonal migrations across ocean basins to track down prey. Lithe and supremely elegant, they glide on fast-moving currents to save energy between meals.

Blue sharks belong to the family Carcharhinidae, or requiem sharks, which also contains the tiger and bull sharks. Named after their eye-catching ultramarine flanks, they are the sleekest, most graceful members of this group, with a very long, rakish snout and scimitar-shaped pectoral fins on either side of a streamlined, gently tapering body. These paired fins act as stabilizers when the sharks are pulled along by powerful currents, as if flying through the water. The longest blues on record measured around 3.8 m (12.5 feet) from nose to tail. However, 2.5–3 m (8–10 feet) is considered an above-average size today, due to the intense fishing pressure driven by the wasteful trade in shark fins destined to be made into bowls of soup. Millions of blues are killed every year to meet this demand, making them the world's most heavily exploited shark.

SHARK TAGGING

Blue shark tagging programmes have proved to be hugely successful at unlocking the secrets of their complex nomadic lifestyle. The distances these fish cover are staggering. One swam over 6,000 km (3,700 miles) from New York to Brazil, and another tagged off the Californian coast was recaught a year later close to New Zealand. Recoveries of tagged individuals provide valuable data about how the sharks use the ocean, allowing conservationists to identify important areas to protect at different seasons, for instance by closing shark fisheries when mating and pupping are taking place.

MARINE HOTSPOTS

The deep-water domain of blue sharks is akin to a marine desert. It is an immense but sparsely populated world of meagre pickings, in which there is very little prey, concentrated in a few hotspots scattered across the ocean. These pinpricks in the seemingly featureless expanse of blue water, barely detectable at the surface, are of enormous importance to marine ecosystems, much as lush oases are in terrestrial deserts. They include, for example, the collision zones where warm and cold currents mingle; the submarine canyons that slice into the shallow continental shelf by the coast; and the nutrient-rich upwellings that bubble up at seamounts (submerged mountains rising from the seafloor).

Not only do blue sharks have a phenomenal ability to locate these productive parts of the ocean, they also turn up at precisely the times of year when food is plentiful. How they manage this is a mystery, although blues, like other sharks, are highly intelligent fish, so it may be possible that they possess a kind of mental map of the ocean. Their extremely acute sense of smell could also serve as a guidance system, by allowing them continually to sample the water en route and thus to follow a chemical gradient of scents dissolved in the water. They dive frequently to hunt squid, going as far down as 350 m (1,150 feet), and travelling through the water column in this way may help them to use their electromagnetic sense to pick up magnetic bearings.

Blue sharks have a truly cosmopolitan distribution, occurring in temperate and tropical waters worldwide, mainly in the open ocean. In the Pacific, they are commonest between 20°N and 50°N, and their numbers undergo marked seasonal variations because many migrate south in winter and north in summer. The populations are sexually segregated, with adult females more abundant further north than males. Similar distribution patterns have been observed in blue sharks of the Atlantic, but the species' movements in the Southern Hemisphere are poorly understood. In the tropics, some populations are thought to be more or less resident throughout the year, although even these sharks may make occasional long journeys of hundreds of kilometres in search of food or mates.

Left Dart-style tags are inserted into a shark's first dorsal fin; they usually carry return instructions in several languages.

Above Blue sharks are known to attack people, but it is they that have most to fear, as the demand for shark fins to supply Asian food markets is decimating their populations.

GOING WITH THE FLOW

Female blue sharks in the North Atlantic carry out a migratory circuit of the ocean in clockwise fashion, which is believed to take 12–25 months from beginning to end (the males, by contrast, do not appear to follow such a clearly defined migration route). The females' amazing journey is up to 15,000 km (9,500 miles) long. It begins off the northeast coast of the USA, where blue sharks gather to mate from around May to July. From this location, the pregnant females ride the North Atlantic Gyre, travelling gradually eastwards in the Gulf Stream and then the North Atlantic Current towards the coasts of western Europe. They reach their main pupping grounds off Portugal and Spain during the following summer, each of them producing a litter of 25 to 50 live young.

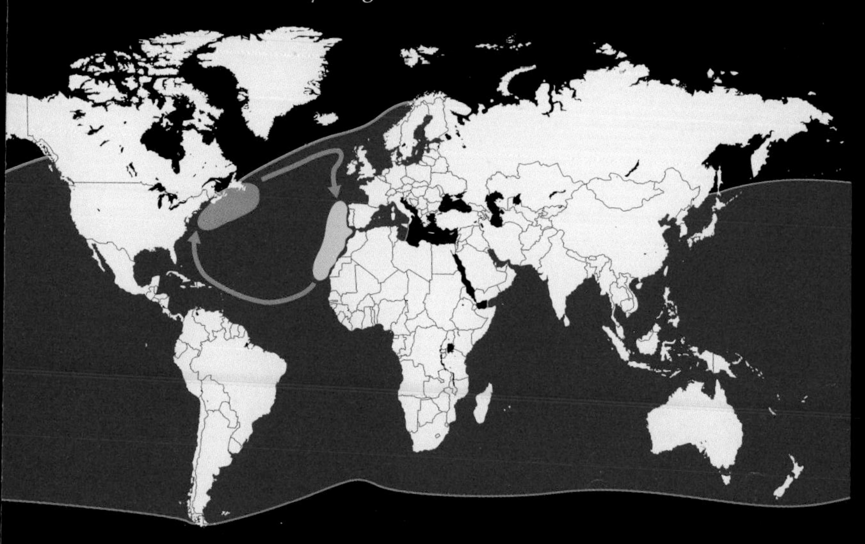

Next the females cruise south along the seaboard of West Africa before using the Atlantic North Equatorial Current to return west to the Caribbean. The last leg is northward up the USA's East Coast, helped again by the Gulf Stream. Since female blue sharks are sexually mature at 5–6 years, live 30 years, and breed every 3 years or so (taking a year off between giving birth and mating), it is theoretically possible that an individual may swim the migratory loop as many as eight times. In practice, some sharks probably remain on the eastern side of the ocean after pupping; many others will tackle the full route only once or twice.

MIGRATION PROFILE	
Scientific name	*Prionace glauca*
Migration	Females migrate between mating and pupping areas
Journey length	Up to 15,000 km (9,500 miles) every 1–3 years (North Atlantic females)
Where to watch	Coasts of southern California, New England, and southwest UK
When to go	October–November (California); May–July (New England); July–August (UK)

BLUE SHARK MIGRATION

Total range **Mating area** **Pupping area**

Eastward migration: adult females after mating
Westward migration: adult females after pupping, and juveniles

Northern Bluefin Tuna

Among the fastest of all marine animals, northern bluefin tuna are designed for endurance swimming. Packed with oxygen-storing red muscle, these warm-blooded superfish are unaffected by cold water and regularly cross the Atlantic as they cruise between distant feeding and breeding grounds.

Tuna are the embodiment of underwater power, with their smooth, torpedo-shaped profile and streamlined fins that retract into grooves and lie flat when moving at speed. All eight species in the genus *Thunnus* share this body plan, but northern bluefins of the northern and central Atlantic are the biggest and most impressive tuna. They surge forwards through the water at speeds of up to 72 kph (45 mph) in short bursts, and cruise at about 48 kph (30 mph). The largest individual on record, caught in 1979, weighed a massive 680 kg (1,500 pounds) and was over 4.25 m (14 feet) in length, although due to overfishing, few if any tuna are likely to attain such a size today.

WARM-BLOODED FISH

Tuna, in common with many swift-swimming sharks (see pages 96–97), have deep pink or red flesh, rather than the greyish or white flesh typical of most fish. The dark coloration is due to their high proportion of red muscle, which is full of blood vessels. Because red muscle cells are supplied with oxygen-rich blood, tuna can metabolize aerobically, and thus produce energy and heat far more efficiently than fish with mainly white muscle; the blood supply to white muscle is poor by comparison, so those fish must make do with anaerobic metabolism.

Moreover, tuna retain metabolic heat much better than average fish. They can keep their body temperature higher than the surrounding water – in other words, they are the closest fish come to being warm-blooded. This means they thrive across an extremely broad thermal range and cope particularly well with cold conditions. Northern bluefins, for instance, occur from equatorial waters to the near-freezing subpolar seas off Iceland and Scandinavia, and dive as deep as 1,000 m (3,300 feet). A second advantage of temperature regulation is that their muscles are able to "run hot" to generate extra power for sustained rapid swimming. As a result, tuna can catch a wide range of fast prey, including herring, mackerel, flying fish, and squid, and have almost no enemies of their own. Much like dolphins, these supercharged fish trap their prey by hunting in packs – they herd the panicked shoals into dense formations called baitballs and then rampage through their victims during a violent feeding frenzy.

MIGRATION PROFILE

Scientific name	*Thunnus thynnus*
Migration	From northern feeding areas to southern spawning areas
Journey length	Up to 10,500 km (6,500 miles) each way
Where to watch	Gulf of Mexico; Mediterranean Sea
When to go	April–June

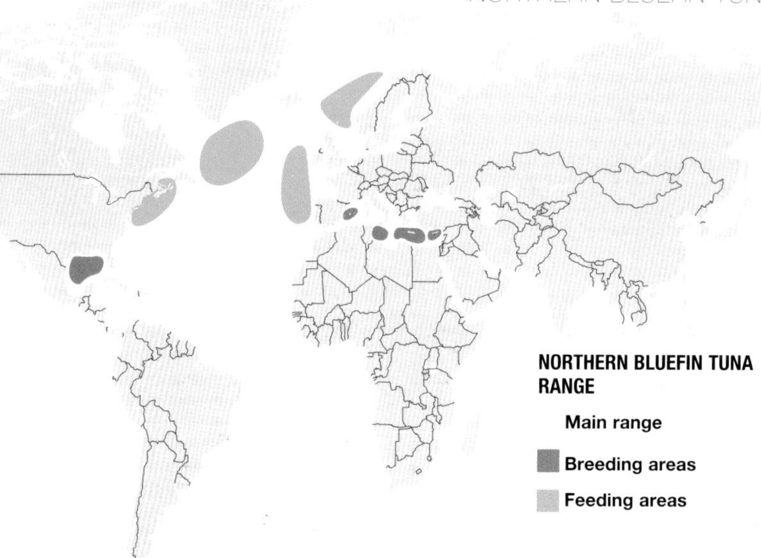

Facing page Northern bluefins are lightning-fast, pack-living predators – truly the wolves of the sea. Their blue backs hide them from prey when they attack from below.

WIDE-RANGING PREDATORS

Tuna inhabit featureless expanses of the big blue, often far from the shallow waters of the continental shelf. It has long been known that they are great wanderers, capable of routine transoceanic journeys, and now satellite tagging programmes have begun to reveal the true scale of their record-breaking migrations. In one such study a tagged Pacific bluefin tuna, *Thunnus orientalis*, traversed the Pacific three times, a total distance of approximately 40,000 km (25,000 miles), in the space of 20 months.

Northern bluefins usually spend at least half of the year in their most productive feeding grounds, and the younger fish stay there for two or three years at a stretch. Important locations include the waters off eastern Canada and the northeast USA, especially an area close to Newfoundland called the Flemish Cap, and the deep waters between Ireland and Spain. Bluefin numbers peak from July to January or February as they gorge on the plentiful shoaling fish, before some of the tuna head southwards to warmer latitudes, where they feed less. It seems that the need to breed is a key factor driving this movement.

Little is known about breeding in northern bluefins. So far two spawning zones have been identified, in the Gulf of Mexico and Mediterranean, where aggregations of sexually mature tuna (those aged 8–10 years or more) take place from April to June. Data gleaned from more than a thousand tagged northern bluefins suggest that these fish are discrete spawning populations, which return to their natal area to breed. However, the fish mingle at other times, moving to opposite sides of the Atlantic. Western-tagged bluefins have been found to undertake transatlantic migrations to reach the Mediterranean, while eastern fish often visit the Flemish Cap feeding area. How the tuna navigate is uncertain; perhaps they use a multidimensional mental map consisting of information about the seafloor, currents, temperature, salinity, and chemical make-up in different parts of the ocean. The fish might also be able to interpret the direction of the Sun's rays and moonlight.

FLOATING GOLDMINES

Industrial fishing has caused a collapse in stocks of northern bluefins, which, together with several other tuna species, are now regarded as critically endangered. Bluefins have been dubbed "floating goldmines" due to the high price they fetch when sold for sushi and sashimi; a heavy specimen can sometimes fetch £50,000 in Japanese wholesale markets.

THE "MATTANZA"

For centuries, as fleets of the largest, mature bluefin tuna enter the Mediterranean to breed, teams of fishermen have waited for them. Traditionally the huge fish are caught by being driven towards the coast, where they can be trapped in a corral formed from dense nets, speared, and then hauled, still thrashing wildly, onto small boats. In Sicily, this spectacular annual ritual is called the mattanza, from matar, an old Spanish word that means "to kill". Held in May or the first half of June, the harvest has a near religious significance to the islanders. In recent years, however, the mattanza has died out as stocks of adult fish in the western Mediterranean have declined by almost 90 per cent since 1975. Today it survives in only a few places on the west coast of Sicily.

Left Although it may appear cruel, the annual mattanza is a more sustainable form of fishing than the industrial harvest carried out by factory ships.

Sockeye Salmon

Sockeye salmon turn their home rivers blood red as they surge upstream in their millions to spawn and then die. On the final leg of a life journey that took them into the open Pacific, they may find their way back to the very lake or stream where they hatched.

Salmon are probably the best known of all migratory fish, because their spawning runs are so spectacular and of huge economic and cultural importance. In the past, the great salmon runs of Alaska and Siberia could block entire rivers, causing them to flood their banks, and northern peoples celebrated the miraculous return of the fish each year. Traditional salmon festivals are still held in summer and early autumn in the northwestern corner of North America.

No fewer than nine species of *Oncorhynchus* salmon, including sockeyes, occur in the North Pacific. In contrast, the North Atlantic is home to a single species, which is placed in its own genus, *Salmo*. The complex life history of salmon, described as anadromous, begins and ends in fresh water but is largely played out in the ocean. Salmon have a variety of physiological adaptations to survive the dramatic changes in salinity and temperature, chief among which is their

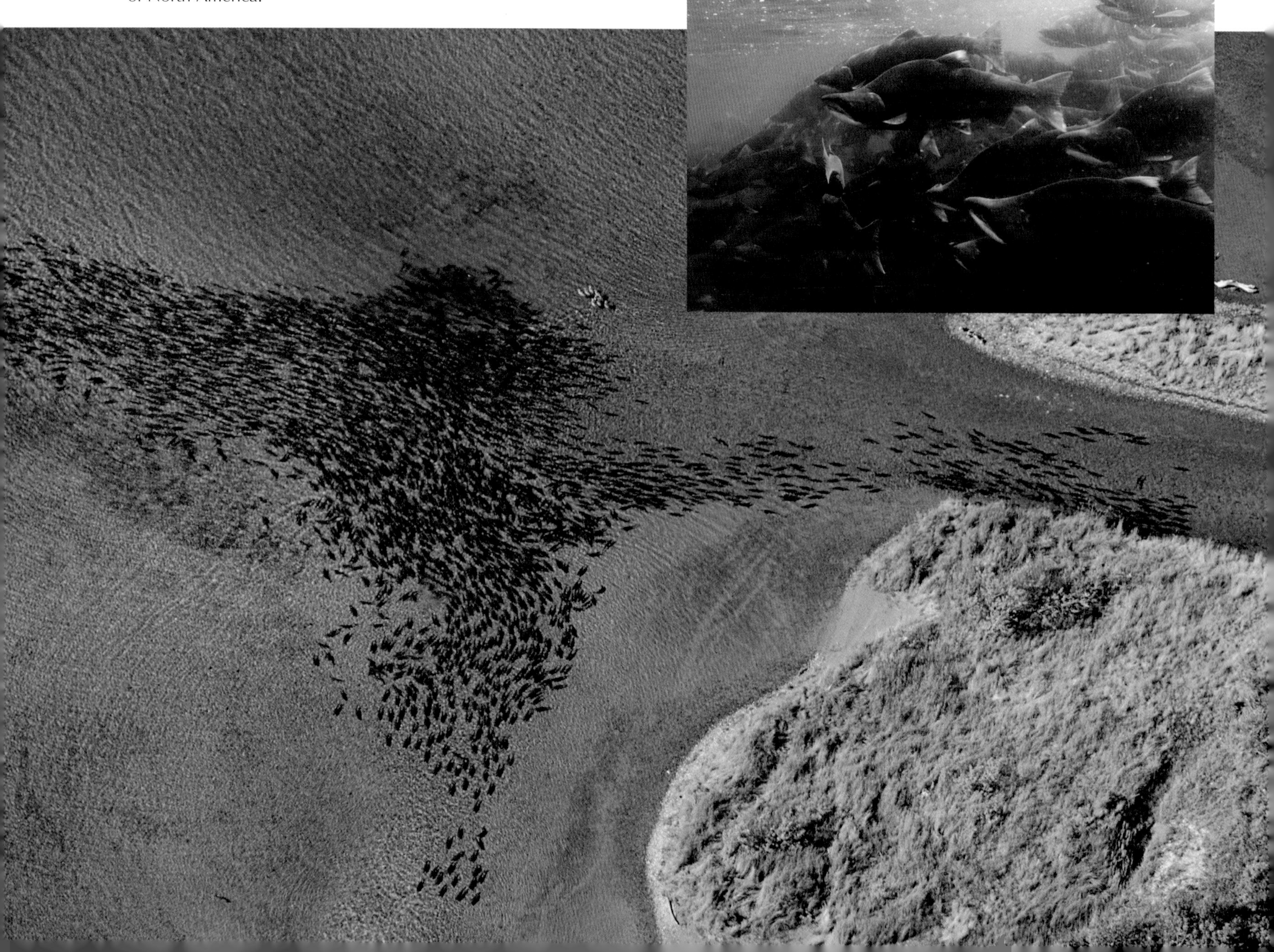

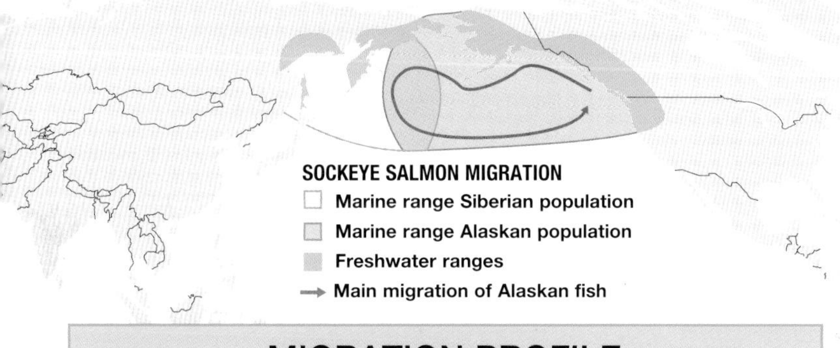

MIGRATION PROFILE

Scientific name	*Oncorhynchus nerka*
Migration	From the sea to lakes and streams to spawn
Journey length	Up to 2,400 km (1,500 miles) in fresh water
Where to watch	Kenai River, Alaska, USA; Adams River, British Columbia, Canada
When to go	June–August (Kenai River); August–October (Adams River)

FISHING BEARS

Salmon form a major part of the diet of Alaskan and Siberian brown bears, which congregate along spawning rivers to feast on the glut of fish. At this time of year, the ever-hungry bears are feeding intensively to put on fat for their winter sleep, and a fully grown adult takes up to 45 kg (100 pounds) of salmon a day. Due to this oily, protein-rich bonanza, the brown bears on Kodiak Island, south of the Alaska Peninsula, are the largest in the world.

Above Bears take years to perfect their fishing skills. Youngsters splash around, but experienced adults like this one wait patiently for stunned or disorientated fish.

kidney function. However, some sockeyes are non-anadromous, meaning that they never leave fresh water. Although these permanent lake residents belong to the same species as migratory sockeyes, they never mix with them.

AGAINST THE FLOW

Sockeyes, in common with their relatives in the order Salmoniformes, have a powerful, torpedo-shaped body and large caudal (tail) fin. This streamlined body plan is suited to endurance swimming. The sockeyes will have travelled several tens of thousands of miles by the time they die at four, or occasionally up to six, years old; further than any other species of salmon. But it is their last journey to breed that is by far the most visible part of their life-cycle and therefore the easiest to study. This one-way pilgrimage from estuary to spawning ground starts in spring or early summer, takes about three to six weeks, and in the largest river systems can be 2,400 km (1,500 miles) long.

As the sockeyes push upriver they wage a non-stop battle against the current and have also to negotiate rapids and falls, leaping clear in a desperate struggle to beat the cascade of water. They do not pause to feed, and instead burn the thick layers of fat stored in their body. Their migration embodies the dictum of "survival of the fittest" – only the strongest make it to the headwaters to pass on their genes. The fish derive their astonishing homing ability from a combination of memory and an acute sense of smell. By sampling the water, which is tainted by the surrounding rocks, soils, and vegetation, they can recognize the unique chemical signature of their native drainage basin.

During the arduous migration, reproductive hormones trigger a grotesque deformation in the male sockeyes. Their jaws stretch into a hook, called the kype, and the once-smooth back becomes humped.

Facing page Masses of sockeyes spawn in the clear shallows of an Alaskan lake. Within a few weeks, these fish will all be dead. **Facing page inset** In their short-lived breeding finery, sockeyes are deep red on the back and sides, with an olive-green head and upper jaw.

In addition, both sexes change colour. Their head turns greenish and their silvery back blushes a beautiful deep crimson, hence the species' alternate name "red" salmon. When the fish at last arrive in their spawning areas, the well-oxygenated waters and coarse gravel bottom will ensure the development of their eggs through the winter. All the spawners expire within a couple of weeks, littering the shallows with corpses. The decomposing carcasses are scavenged by bears, bald eagles, and wolves and return vital nutrients to the freshwater food chain.

Rising water temperatures in April and May trigger hatching of the larval fish, or alevin. These turn into salmon fry, which develop in streams and lakes for one or two years and then head downriver, timing their movement to coincide with the "freshet" – the torrent of spring meltwater. At this stage, the young salmon are known as fingerlings or smolts. Adult sockeyes feed on plankton and small fish and mature at sea for between two and four years, ranging from coasts out into mid-ocean; some fish of Alaskan origin may even turn up in Japanese waters.

LAST OF ITS KIND

The sockeye migration varies according to a predictable pattern – the run that occurs every fourth year dwarfs the others. But overall there has been a steady decline in numbers in recent decades due to the many hazards faced by migrating salmon, including weirs and hydroelectric dams (the provision of fish ladders alleviates this problem without resolving it). Already certain sockeye populations have died out, such as the so-called Redfish Lake salmon in the USA. Prior to European settlement, about 10–15 million sockeyes used to journey up the Columbia and Snake rivers to spawn in Idaho's Redfish Lake. These rivers have since been dammed 11 times to provide cheap power and irrigate the arid Columbia Plateau of eastern Washington state, and by 1992 only a single wild Redfish Lake salmon was left, a male nicknamed "Lonesome Larry".

European Eel

MIGRATION PROFILE

Scientific name	*Anguilla anguilla*
Migration	Between European fresh waters and Atlantic
Journey length	Probably up to 8,000 km (5,000 miles) in salt water
Where to watch	Rivers throughout Europe
When to go	September–November (at night)

European eels are fascinating fish with a dual personality. Having matured in fresh water, they set off one autumn to swim thousands of miles across the Atlantic to spawn. The larvae drift back on ocean currents, a mammoth migration for such tiny animals.

When fully grown, adult European eels can be 1.2 m (4 feet) in length and weigh as much as 6.6 kg (14 pounds), although they typically reach two-thirds this size. These strange, tube-like fish are often mistaken for snakes because of their elongated form and sinuous style of swimming, driven by muscular waves that ripple along the length of their body. The absence of pectoral and pelvic fins sticking out from either side enhances this serpentine appearance, as does the lack of any gill covers, which are reduced to a small opening.

European eels, also called common eels, are one of several closely related species in the family Anguillidae. All have an unusual type of migratory life-cycle, described as catadromous. They begin life as planktonic larvae in the open ocean and years later die there as breeding adults, but inbetween live in fresh waters, sometimes huge distances from their marine nursery grounds. Special adaptations in their kidneys and gut enable them to survive the transformation from salt to fresh water and back again. The eels have just one chance to reproduce, and die soon after spawning in deep water.

LONG DEVELOPMENT

As with many other true eels, or Anguilliformes, these fish are exceptionally long-lived. Males spend up to 10–12 years feeding and growing in fresh water, and females even longer, in some cases up to 20 years; there are unsubstantiated reports of 30-year-old eels. During their freshwater phase, they haunt a wide variety of habitats, from tidal rivers to marshes, pools, drainage ditches, and streams. Surveys show that younger eels tend to be more abundant in the lower reaches of rivers, while older, larger individuals frequent the upper stretches.

By day, the eels lurk among tangled roots and submerged vegetation to avoid their main predators – herons and cormorants. After dark, they emerge to hunt invertebrates and other small prey and to scavenge carrion. Remarkably, they can survive out of water

Left On dark, wet nights in the autumn, adult eels can be spotted sliding over the ground during their migration back to the sea, but most of their journey takes place unseen, underwater in rivers and streams.

for brief periods, due to their leathery, slime-coated skin, and this allows them to slither across the ground in search of prey or to colonize new habitats. In winter, the eels burrow into the soft mud at the bottom of rivers and ponds to sleep out the coldest months in a torpid state.

NOCTURNAL DEPARTURE

Eventually, mature eels are ready to leave their freshwater lairs to make the long journey out to sea to breed. The migration peaks on dark nights in the autumn, especially between the last quarter and new moon. The cue for the eels' departure is a drop in water temperature and the sudden increase in water flow caused by heavy autumn rains. Helped by the current, the eels can cover 50 km (30 miles) a night, moving in dense groups.

As the eels near the coast, their gut shrinks and they cease to feed, and their eyes nearly double in size, suggesting that they migrate at depth in the ocean. In addition, their dark bodies turn silver, probably for camouflage in open waters. Very little is known about the spawning behaviour of European eels, which has never been observed in the wild. However, they are known to produce immense quantities of microscopic eggs that float with the current, finally hatching into transparent larvae, called leptocephalae. Near-invisibility is a good defence against predators such as fish, but even so, only a tiny fraction of the eel larvae will finally arrive safely in river mouths, after two or three years adrift. While they are there, the larvae morph into pencil-sized young eels – elvers – and then head upstream like a mass of wriggling spaghetti to continue the species' life-cycle.

SPAWNING SECRET

The spawning grounds of European eels have yet to be discovered, but are thought to lie in the Sargasso Sea in the western Atlantic, east of Bermuda. From here, the larvae would be able to return to Europe by riding first the powerful Gulf Stream current, which transports warm, salty water from the Caribbean in a northerly direction, and then its northwest continuation, the North Atlantic Drift, which flows past the seaboard of western Europe.

Above Elvers have such transparent bodies that you could read the pages of this book through them. Yet, despite their fragile appearance, they endure a huge swim upriver.

An international research programme is underway to establish if this is the case. By attaching satellite tags to migrating "silver" eels in autumn, scientists hope to collect data on the route taken by the eels while at sea, and perhaps to pinpoint their final destination. A separate study, in which captive eels were placed in a swim tunnel, has already proved that the fish are super-efficient swimmers capable of completing a simulated migration of 5,500 km (3,400 miles) without feeding. Further tests may be needed, because the distance from the Sargasso Sea to northernmost parts of the species' range is even further, at around 8,000 km (5,000 miles).

MYSTERIOUS DECLINE

The days when every major European estuary boiled with vast shoals of migrating juvenile eels are over. Numbers of elvers reaching the continent's shores have fallen by at least 90 per cent since the 1970s. Some authorities have argued that this decline is natural in a cyclical population, so will be short-lived, but most believe it represents a long-term, possibly irreversible, trend. Overfishing is assumed to be a contributory factor. More serious threats may include pollution; barriers to migration, such as tidal barrages and hydroelectric dams; and nematode worm infestation.

Above Within a human generation, the sight of a river overflowing with migrating European eels has become a rarity.

The Greatest Migration

After dark the greatest mass migration on Earth takes place in the open ocean. Billions of animals, from minuscule crustaceans to jellyfish, squid, and sharks, rise through the water column to the surface, then sink back down again at dawn.

The top 30 m (100 feet) of the ocean teems with phytoplankton – microscopic bacteria and blue-green algae that harness the Sun's energy to manufacture food. Virtually all marine life ultimately depends on the phytoplankton. Its primary consumers are immense swarms of tiny animals, including single-celled protozoa, shrimp-like copepods and krill, and strange, transparent, tube-shaped organisms called salps. This ragtag army, known collectively as zooplankton, spends the day resting in deeper waters, where it waits until dark before before ascending to graze.

The nocturnal journey of herbivorous zooplankton to surface waters is the basis of a phenomenal migration involving a host of other species. Following in their wake are small carnivorous zooplankton: arrow worms; shrimps; sea butterflies (also called flapping snails); comb jellies; and the free-swimming larvae of countless jellyfish, crabs, and fish. These creatures in turn attract larger members of the zooplankton, such as shoals of lanternfish and squid, which emit a cold light known as bioluminescence, possibly to find or lure their prey. Next in the food chain are dolphins, turtles, and sharks. The largest migrants are monsters: ferocious jumbo squid up to 2 m (7 feet) long; and slow-moving megamouth sharks, which reach at least 5.5 m (18 feet) in length.

Scientists do not fully understand the driving force behind this vertical migration. However, zooplankton probably retreat to deep water to avoid surface predators active in daytime and perhaps because they use up less energy at cooler depths.

Left Twice every 24 hours, at sundown and again at first light, an entire marine community is on the move. Jellyfish are among the most numerous of these vertical migrants.

Below left Squid are highly mobile nocturnal hunters that rise through the water column in dense packs. This photograph shows a pelagic squid of the genus *Chiroteuthis*.

Below centre Hatchetfish, in common with many open-ocean predators, usually attack from below. Their fixed upward gaze helps them to discern the movements of prey above.

Below right Tiny zooplankton, such as these copepods and crab larvae, are the world's smallest migratory animals. Without them, the seas would be lifeless deserts.

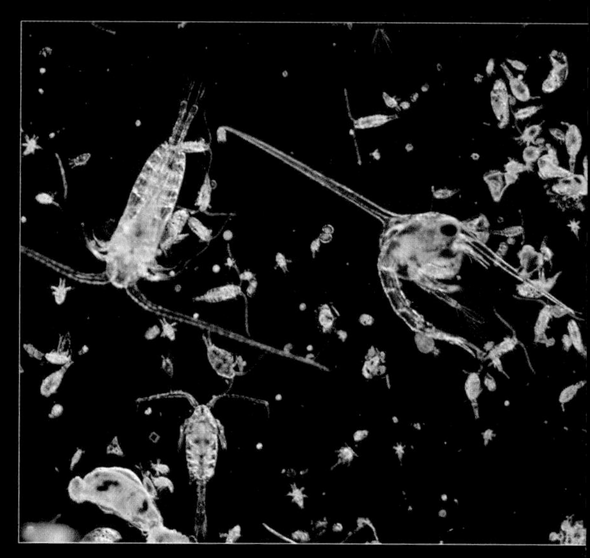

ATLANTIC KRILL RANGE

Total range

Above The transparent outer shell, compound eyes, and oar-like swimming legs of krill are clearly visible in this macro photograph.

MIGRATION PROFILE

Scientific name	*Euphausia superba*
Migration	Daily vertical movement to surface waters in summer; seasonal movements between sea ice and open ocean
Where to watch	Antarctic waters
When to go	December–February

WHALE FOOD

Krill is an old Norwegian word meaning "whale food" – an apt description. Many whales feast on krill, while blue whales eat nothing else. A fully grown blue whale weighs up to 200 tons and to sustain its bulk needs 1.5 million calories a day. But since it breeds in warm, tropical waters almost empty of krill, the whale can feed only during the few months of the year it spends in polar seas; therefore, it consumes at least 3 million calories daily – equivalent to roughly 4 tons of krill. Like other baleen whales, the blue whale feeds by taking huge gulps of seawater and sieving out the edible contents using the baleen plates on its upper jaw. One species of Antarctic seal, the misleadingly named crabeater, also specializes in feeding on krill, and has evolved interlocking teeth to strain them from the water. By exploiting this prolific resource, the crabeater seal has become one of the most numerous of all large mammals, with a population 15 million strong.

Antarctic Krill

Antarctic krill swim through the Southern Ocean in uncountable millions, at times staining the surface waters red. Enormous swarms of these shrimp-like crustaceans follow the annual melt and freeze of the sea ice, and sustain a host of other animals, from penguins to whales.

If not for their shoaling habits, krill would be virtually invisible to the naked eye, as these creatures are transparent and a mere 6–60 mm (0.25–2.5 inches) long. In fact, the largest shoals are visible from space and can be tracked by satellite. There are at least 89 species of krill, found worldwide as part of the plankton, but the highest densities occur in cold temperate and polar seas. Of the dozen or so species in the Southern Ocean, Antarctic krill are by far the most abundant: one swarm covered an area of 450 sq km (175 square miles) of ocean, to a depth of 200 m (650 feet), and was estimated to contain over 2 million tons of krill.

CLOUDS OF KRILL

Shoaling Antarctic krill make up the greatest concentrations of animal biomass ever recorded, so it is no surprise that krill are the most important source of protein in the Southern Ocean. They are the foundation of Antarctic food chains, being consumed in immense quantities by all higher life forms, including fish, squids, jellyfish, penguins, seals, and baleen whales.

The krill live for around five years and every winter migrate down under the pack ice. It seems that the adult krill barely feed throughout the six-month-long polar winter, and survive by using up fat stores laid down during the previous summer; they slow their metabolism and shrink in size, regressing to a kind of juvenile-like form and even eating their own shell to stay alive. Beneath the ice there are also genuine immature krill, for which the pitch-black, icy tomb is a safe nursery ground with few predators.

As sea temperatures rise in summer, the krill migrate back to surface waters along Antarctica's shoreline, freshly exposed by the retreating pack ice. Here the long hours of daylight cause a sudden bloom of phytoplankton (microscopic plants), especially single-celled algae known as diatoms. This thick "soup" is what the krill have come to harvest. Their modified front limbs busily filter the diatoms from the water, together with smaller amounts of zooplankton (tiny animals such as fish, mollusc, and jellyfish larvae).

By midsummer, the pack ice has disappeared completely and the swarms of krill reach their greatest extent, often many miles across. For safety, every 12 hours the krill sink deeper, beyond reach of most predators, rising to the ocean's upper layers again 12 hours later. They are efficient swimmers, beating their rows of oar-like legs to move through the water column. The daily migratory cycle is repeated over and over until the ice starts to re-form with the onset of winter, forcing the swarms to disperse.

MIXING CURRENTS

Swarms of krill tend to occur patchily, because their distribution is linked to nutrient-carrying currents and upwellings. Indeed, recent research suggests that the krill do not simply drift passively in these currents but actually modify them. By moving vertically through the ocean on a 12-hour cycle, the swarms play a major part in mixing deeper, nutrient-rich water with nutrient-poor water at the surface. Unfortunately, ocean currents are affected by climate change, and so global warming could have a disastrous impact on both krill and the wider Antarctic ecosystem.

Conservationists are also alarmed by the Southern Ocean's fast-developing krill fishery. In 2008, the annual krill harvest reached 800,000 tons. Most of this krill is turned into fish-farm feed or health supplements, such as Omega 3 oil. Despite evidence that populations of Antarctic krill have declined by up to 80 per cent since the 1970s, there are plans to increase the krill catch using larger factory ships and new "suction" fishing technology.

Below left In early summer adult krill lay eggs, which sink about 300–400 m (1,000–1,300 feet) to the continental shelf. The larvae pass through several stages as they swim to the surface and spend the winter grazing algae under the sea ice. Larvae of eggs that sink beyond the shelf may starve before they make it back to the surface. **Below right** Krill are a keystone species, providing food for humpback whales and many other summer migrants.

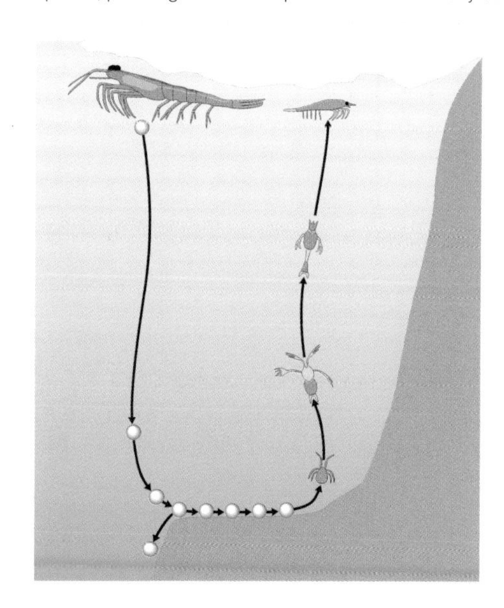

Caribbean Spiny Lobster

At the first hint of coming winter storms, Caribbean spiny lobsters link together for a mass conga across the seabed to the sanctuary of deep water. Guided by a magnetic map, they follow particular routes and have an extraordinary capacity to navigate home again in spring.

Many of the world's decapods – the order of ten-legged crustaceans that includes lobsters, crabs, and shrimps – are known migrants. Tag and recapture studies demonstrate that they often move between separate feeding, breeding, and wintering ranges. Spider crabs also travel to communal moulting areas, forming huge mounds to shed and replace their carapace in safety. But Caribbean spiny lobsters are unique in that their seasonal movements have been watched and filmed underwater. Indeed, there is now ample evidence that spiny lobsters possess navigation abilities every bit as sophisticated as in vertebrates, despite having a much simpler nervous system.

Caribbean spiny lobsters grow up to 60 cm (2 feet) long and are native to the Gulf of Mexico, Caribbean, and western Atlantic, from North Carolina and Bermuda south to Brazilian waters. They lack the massive claws of the much larger European and American (or Atlantic) lobsters, which belong to the cold-water genus *Homarus*, but the vicious spines that cover their carapace and antennae deter most predators. In summer they inhabit warm shallows, especially over coral reefs and around mangrove islands, hiding in crevices by day and emerging at night to hunt for invertebrate prey.

Each spiny lobster has a foraging range of perhaps a few hundred square yards, within which it uses a number of daytime hide-outs. It may stick to the same patch for several months, or move on after only a few weeks; either way, an individual will gradually acquire a detailed picture of the seabed over a wide area during the course of an average 15- or 20-year lifespan.

SEAWARD MARCH

Young spiny lobsters spend their first three or four years in nursery areas (usually seagrass meadows), whereas the adults undergo a dramatic behavioural change in autumn, abandoning their territories to assemble in large, restless groups that are active by both day and night. These lobsters form chains of as many as 50–60 individuals and march in single file along the sandy ocean floor towards deeper water. Although they are reasonably fast swimmers, they clearly prefer to travel on foot, probably because they can maintain close physical contact with their antennae and bunch together to defend themselves in the event of an attack by a shark, grouper, or some other predator. The swaggering lobster trains cover up to 14 km (9 miles) in a day.

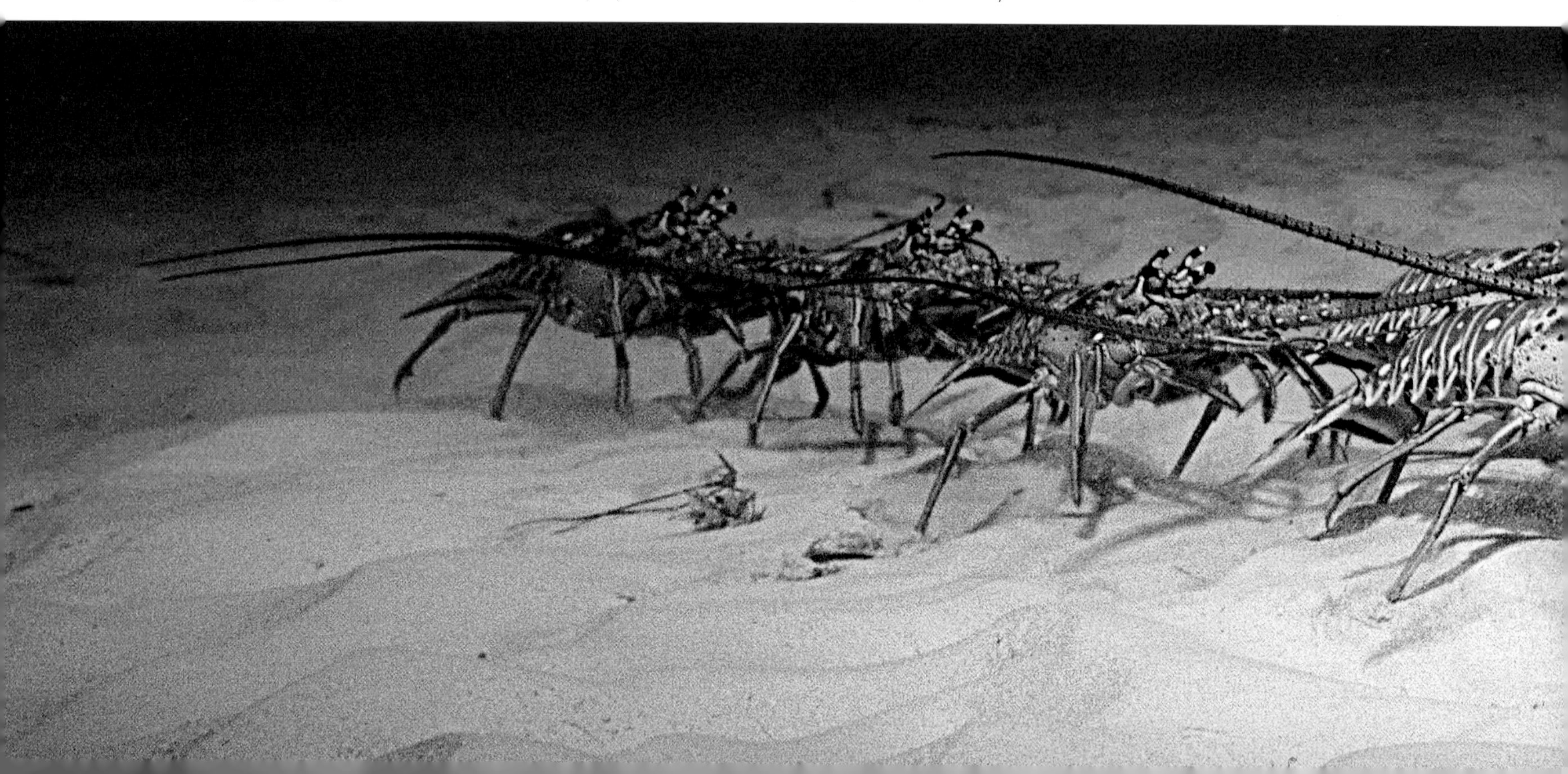

CARIBBEAN SPINY LOBSTER MIGRATION

Total range → Seaward migration → Landward migration

MIGRATION PROFILE

Scientific name	*Panulirus argus*
Migration	To deeper water in winter
Journey length	Up to 50 km (30 miles) each way
Where to watch	Coastal waters off Florida Keys, USA
When to go	April–July

It seems that the lobsters' annual trek is triggered by shorter days or the sharp fall in water temperature associated with the first serious storm of autumn, or maybe these cues work in combination. Another trigger might be the increasing turbulence and cloudiness of seawater at this time of year. The winter sojourn on deep-water reefs enables lobsters to avoid a buffeting from storms tearing through the shallows; they save energy too, because in cooler water they can slow their metabolism, so in turn do not need to feed as often. Retreat to deeper water may therefore be a method of waiting out the lean winter months.

MAGNETIC ROUTE-FINDING

We have long known, from the multiple recaptures of tagged individuals, that Caribbean spiny lobsters are able to find their way back to the same foraging grounds in summer, and even to the same dens among the coral. They orientate by sight, using their knowledge of the underwater terrain, and by comparing their position to the wave surge – the horizontal movement of water near to the sea floor. However, with thousands of lobster groups on the move, all following nearly identical compass bearings, sometimes in total darkness, there must be another, more accurate navigational system at work as well.

Spiny lobsters were thought to orientate using a magnetic compass sense, deriving positional coordinates from Earth's magnetic fields, but this was not proved until the 1990s. In one experiment, magnetic coils were buried in the sand and the polarity of the coils then switched, to see what effect it had on lobsters tethered nearby. Observers noted that the lobsters changed course in response to changes in the magnetic field. In 2003, spiny lobsters were proved to use their magnetic compass not just for direction but to locate their precise point on Earth – the first time this had been shown in any invertebrate.

VALUABLE HARVEST

Caribbean spiny lobsters mature slowly – they do not breed until about five or six years old – so like most large lobsters around the world are at risk of overexploitation. They are caught by recreational divers and harvested commercially. In the past spiny lobster catches slumped due to intense fishing pressure, but the Gulf of Mexico fishery has been tightly controlled since the 1970s to avert a future collapse.

Above left Individual lobsters have the ability to relocate the same den from one year to the next. **Below** Lobster chain gangs are a practical solution to the problem of how to migrate across open areas of seabed without getting attacked by predators.

Australian Giant Cuttlefish

Giant cuttlefish converge on the south Australian coast each autumn to breed, hovering over the rocky reefs in rippling, kaleidoscopic shoals. Within weeks they float to the surface dead, completely spent after their spawning frenzy. Large flocks of seabirds arrive to feast on the tide of lifeless bodies.

AUSTRALIAN CUTTLEFISH MIGRATION

- Spencer Gulf breeding hotspot
- Breeding range
- Migratory range
- → Migration movement

MIGRATION PROFILE	
Scientific name	*Sepia apama*
Migration	To spawning grounds off southern Australia
Journey length	Not known
Where to watch	Spencer Gulf, South Australia
When to go	May–September

Facing page Rival males spar energetically on the reef, producing pulsating stripes of colour as they attempt to overawe their opponents. **Right** The cuttlefish can change appearance in the blink of an eye – a useful trick for ambushing prey and hiding from their own enemies.

More than a thousand species of squids, octopuses, and cuttlefish have been recorded so far – some along coasts, some in surface waters, and others in the inky depths – and new species are identified every year. Known collectively as cephalopods, these ancient molluscs are today the most highly evolved marine invertebrates, with big brains and complex behaviour. Many are strong migrants, moving vertically through different layers of the ocean every 24 hours, and they may travel great distances between feeding and spawning areas.

Of all cephalopods, Australian giant cuttlefish carry out arguably the most spectacular, easily observed migration. It takes place just offshore in crystal-clear shallows, and involves hundreds of thousands of squabbling cuttlefish, all desperate to seize their last opportunity to reproduce before they die. The drama unfolds during the austral (Southern Hemisphere) autumn, from May to July, with the peak season depending on location and sea temperature. Unusually for members of the typically nocturnal class Cephalopoda, giant cuttlefish remain active by day when spawning. This has helped ecotourism to thrive at prime sites, such as Spencer Gulf in South Australia, where it is possible to dive or snorkel among the dense clouds of cuttlefish.

SHIMMERING SHOALS

Giant cuttlefish live for no more than three years; nevertheless, true to their name, they may reach an impressive 1.5 m (5 feet) in length and weigh up to 14 kg (30 pounds). Little is known about their life history, especially during the first couple of years spent out at sea. However, their magnificent breeding rituals have been documented extensively.

Male giant cuttlefish, which are larger than the females, intimidate rivals by displaying furiously at each other. They hover in the water in their dozens, like a bizarre fleet of spacecraft from a science-fiction film, while producing split-second pulses of brilliant colour that flicker over their entire body. "Zebra" stripe patterns flash across their soft skin in mesmerizing combinations of iridescent reddish-purple, yellow, and greenish-blue, sometimes turning dark and at other times ghostly pale. This stunning light show is made possible by iridophores – cells in the lower layer of cuttlefish skin that reflect polarized light. Ordinarily, these cells are used to help the cuttlefish blend in with their surroundings, but now they are deployed by the males in a form of high-intensity visual warfare.

The competing males jockey for position, gliding forwards, backwards, and sideways in a graceful underwater ballet as they all try to defend a portion of the reef. Eventually the dominant males start displaying to the watching females, which tend to stay close to the reef, among seaweed or rocks. After mating, each female cuttlefish lays her eggs in crevices in the reef, and the species' life-cycle is complete.

UNSOLVED MYSTERIES

The purpose of the giant cuttlefish migration is clear enough: the cuttlefish need a hard substrate to lay their eggs and warm water for them to develop, and only shallow rocky reefs meet both these requirements. But how do the widely dispersed cuttlefish know when to set off towards the coast? Several possible triggers have been suggested, including changes in the light, temperature, water density, and salinity, although evidence for how these factors might work is lacking.

Like most cephalopods, giant cuttlefish are efficient swimmers, moving with undulating movements of their lateral fins, which resemble a frilly "skirt". Some squids, such as the Atlantic short-finned squid, have been shown to cruise thousands of miles with ocean currents, so there is no reason to suppose that giant cuttlefish cannot also drift long distances. However, how they find their way is unclear. Perhaps the cuttlefish orientate using information about currents, or maybe they use chemical clues, sampling the water to check progress. One possibility is that they possess a geo-magnetic sense, enabling them to establish their direction relative to Earth's magnetic fields.

A persuasive theory is that their superb vision has a vital part to play. Giant cuttlefish have huge compound eyes, and much of their brain is devoted to processing visual data. Although they see only shades of black and white, they can perceive plane-polarized light. Light passing through the atmosphere is diffused and reflected, so that it vibrates in one plane and is said to be polarized; as the Sun's position changes throughout the day, the overall pattern of polarized light changes too. The cuttlefish might be able to use these shifting, solar-generated patterns to navigate.

SEABIRD BONANZA

At the end of each year's spawning, the slicks of dead and dying cuttlefish attract ocean-going birds. Many are doubtless opportunists, alerted by the stench while passing through the area, but others appear to route their migrations specifically to take advantage of this annual bonanza. Individual birds may even turn up at the same inshore reefs year after year to feast on cuttlefish.

Above Soaring effortlessly on stiff wings, albatrosses arrive to gorge on the abundant dead and dying cuttlefish.

Above Waterfowl are powerful migrants. Waves of Canada geese push south from their tundra nesting grounds in the autumn, using traditional flyways to reach milder climes.

Migration by Air

From the tropics to the poles, the Earth's skies are criss-crossed by countless migration routes, which at peak times become as congested as any urban expressway. Birds are the best-known airborne voyagers – about half of all species of bird migrate regularly. Some travel only a few miles, up or down mountains for example, but others circumnavigate the globe, or disperse far and wide without a fixed itinerary. The other great aviators are bats and insects. In parts of the Americas and Africa, it is possible to see clouds of migrating butterflies or dragonflies several miles long.

MIGRATION PROFILE

Scientific name	*Tadarida brasiliensis*
Migration	Part of US population migrates south in winter
Journey length	Up to 1,800 km (1,100 miles) each way
Where to watch	Carlsbad Cavern, Texas; Congress Avenue Bridge, Austin, Texas
When to go	August–September

Above Mexican free-tails are highly capable migrants that can fly both fast and high, breaking their journey to sleep in trees and caves. Here, high-speed flash captures one of the bats leaving its daytime tree-hole roost.

Mexican Free-tailed Bat

Free-tailed bats are the long-haul champions of the bat world. Each year millions fly from Mexico to form megacolonies in the southwest USA, and at sundown the bats whirl out of their giant roosts in dense, spinning clouds that resemble plumes of smoke.

Chilly winter air lacks enough food to sustain insectivorous bats, so in temperate regions of the world they hibernate until spring. At one time naturalists assumed that all temperate bat species did this: it was only in the early 20th century that people began to suspect bats might also be migrants. However, proof was not forthcoming until the mid-1950s, when ringing studies demonstrated beyond doubt that individual bats could travel far from their original roost.

It is now known that migratory behaviour is widespread in the bats of North America, Europe, and Asia. Some species move relatively short

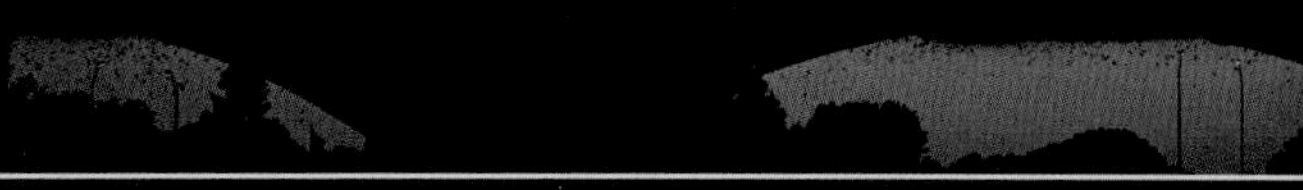

Right Watched by a crowd of spectators, thousands of free-tailed bats stream from beneath Congress Avenue Bridge in Austin, Texas. Their huge colony makes its home in crevices beneath the bridge.

distances of under 80 km (50 miles), but a handful regularly migrate 1,600 km (1,000 miles) or occasionally even further, with the speed and navigational skill to rival birds. The Mexican free-tail is the best documented of these long-distance migrants.

RESIDENT AND MIGRATORY

In Mexican free-tails, like all members of the family Molossidae, the lower half of the tail is free of the membrane that joins the base of the tail to the hind legs. The first part of this species' name is less apt, because the bats occur in a wide area that stretches from southern Oregon east to Kansas in the USA, then south through Texas and Mexico to Central America and northern South America. Across this range, some of the bat populations remain in the same district all year round and others are strongly migratory.

Within Central and South America, Mexican free-tails are believed to be mostly sedentary. In parts of the USA, too, these bats are non-migratory; those from Oregon and California, for instance, live in a region of typically mild winters and are able to survive by sheltering in warm buildings or entering a torpor for short periods. But the free-tails that breed elsewhere in the USA, including Texas, where the biggest numbers are found, visit only during the spring and summer months. They arrive in late February and fly south again between late October and early November. Some travel to western Mexico, but the majority of them head farther south and east to Tamaulipas, Coahuila, and Nuevo León.

The free-tails often stagger their southbound migration by stopping off en route, and their journey may involve hops between different bat caves. They are powerful fliers, however, so can travel rapidly if necessary. Their long, narrow, angular wings give them a distinctive rakish silhouette and a fast-paced flight style as they chase moths during nocturnal foraging trips, sometimes climbing to impressive altitudes. Bats from Bracken Cave near San Antonio in Texas have been tracked with Doppler radar, leading to the startling revelation that in June they fly as high as 3,000 m (10,000 feet) to prey on recently emerged swarms of corn earworm (cotton bollworm) moths drifting with the prevailing wind. If the bats are capable of high-level feeding flights, then it is possible that they use this ability during their migrations, to cruise with strong tailwinds at altitudes of hundreds or thousands of feet. Since long-distance flight requires a lot of energy, any behaviour that enables bats to cover ground faster could be vitally important.

MATERNITY COLONIES

The night-time migrations of Mexican free-tails are extremely difficult to observe, so it is not surprising that they are famous primarily for forming enormous maternity colonies – the largest gatherings of any warm-blooded animal. The biggest known colony, formed by almost 20 million bats, is the Bracken Cave colony, outside San Antonio, Texas; in the past, before pesticide poisoning and deliberate persecution caused bat populations throughout North America to crash, there were other so-called "guano caves" that contained similarly sized colonies.

Female Mexican free-tails raise their single pups entirely alone, and as a result the maternity colonies normally contain few males, which sleep in separate "bachelor" roosts nearby. The bat population that visits Texas in summer is most unusual in that it has a particularly small proportion of mature males, which are not very sexually active. It seems that many of the males from this population do not visit Texas as adults and instead remain in Mexico, and that they mate with the migratory females before they fly north each spring.

URBAN BATS

Free-tailed bats have experienced mixed fortunes in the USA. Long persecuted out of superstition and fear, and for carrying rabies, they disappeared from many former haunts after their nursery sites were blocked up or vandalized. Today, the bats are benefitting from a more enlightened attitude. Bat watching is growing in popularity – not just at well-established sites such as Carlsbad Cavern in Texas, but also in some unexpected places. One of the world's best locations to watch free-tails is downtown Austin, where there is a thriving colony on Congress Avenue Bridge. An estimated 1.5 million bats spend the summer on crevices under the bridge, and bat tourism now contributes £5 million to Austin's economy per year. Other large urban free-tail colonies are at Waugh Street Bridge in Houston and Yolo Causeway near Sacramento, California, while several American cities have incorporated "bat-friendly" features into new civil engineering projects.

Straw-coloured Fruit Bat

Straw-coloured fruit bats migrate across tropical Africa in their millions, moving unseen through the night skies in secret swarms. Bats from many different regions travel hundreds of miles to the same temporary roost, from where they fly nightly sorties to gorge on the local fruit crop.

Named after the soft, golden fur on its back and shoulders, the straw-coloured fruit bat is easily the biggest bat in mainland Africa, with an average wingspan of 80 cm (30 inches). It is a heavy bat built for stamina rather than aerobatic ability – the first clue to its migratory lifestyle. Long, broad wings give the bat a high wing loading (the ratio of wing area to body mass), which is ideal for long-distance cruising but not for precision manoeuvres. As a result it can feed only from the forest canopy and the outer branches of isolated trees.

The straw-coloured fruit bat breeds throughout equatorial Africa, in smelly treetop colonies or "camps" that contain thousands of individuals. Often these are located in city streets or near waterfalls, raising the interesting possibility that noise may in some way help breeding success. The colonies are occupied for most of the year, then abandoned for about three months for no apparent reason. A three-year study of a huge outdoor roost at Kampala, Uganda, showed that bat numbers here plummeted to fewer than 10,000 from a previous maximum of 210,000. Clearly, the bats are travelling away from the region en masse, but where they go and why has long been a puzzle.

FRUIT BURST

According to research carried out by American biologist Heidi Richter in Zambia's Kasanka National Park, the answer to the riddle of the disappearing bats could be the explosion of fruit production in a localized area outside the bats' usual foraging grounds. She has likened this phenomenon to a kind of ecological "Big Bang". Lots of trees at Kasanka, including many species of *Syzygium* and *Uapaca*, bear fruit between September and December during the wet season. This harvest is much too large for resident fruit bats and other animals in the park to

Above The migration of straw-coloured fruit bats is driven by seasonal fruiting patterns across Africa, which in turn depend on the annual rains.

Inset By day, roosting bats gather in high trees, where they socialize and digest last night's feast.

GUARDIANS OF THE FOREST

Fruit bats occur only in the tropics of Africa, Asia, and Australasia. Together the 175 or so species make up a single suborder: Megachiroptera. Many of them are known to be migratory, but the straw-coloured fruit bat's seasonal movements are probably the most remarkable. Fruit bats play a vital role in forest and savanna ecosystems as pollinators and dispersers of seeds.

cope with, creating a short-lived surplus that attracts migrants from far away. Between five and ten million straw-coloured fruit bats converge on Kasanka each year, briefly forming what is thought to be the largest gathering of mammals on the African continent.

The bats time their visit for when the fruit crop reaches its peak: strong evidence that the sudden abundance of food is the main factor driving their migration. Groups of bats usually begin arriving at Kasanka in late October, and over the next three weeks this trickle turns into a flood. Their departure is more rapid. Virtually all the bats leave the area within the space of one week, either at the end of December or sometimes in the first half of January.

CROWDED ROOSTS

Kasanka's hordes of straw-coloured fruit bats establish their roosts in two short stretches of evergreen swamp forest. At dawn the bats stream back to the roosts from a night's feeding and swoop into the tall mushitu trees, where they settle in tightly packed clusters to sleep. Examination of the roosting bats, which can be captured by hand without difficulty, have shown that some are pregnant females, and that their gestation is at different stages. Since the females in each breeding colony tend to synchronize their births, this suggests that the bats at Kasanka originate from a variety of regions across tropical Africa. The mixing of different breeding populations might be of great importance to the species, because if bats were to find partners and mate at the roost it would widen the gene pool significantly.

To discover where the bats go when they leave Kasanka, a few were fitted with a lightweight satellite transmitter, held in place with a custom-made collar. Although inconclusive, the resulting data proved that at least some of the Kasanka bats fly a very long way. One male covered 1,900 km (1,180 miles) in six months, before his signal stopped in the Democratic Republic of the Congo. Many more straw-coloured fruit bats will need to be tagged to map their migration routes. In addition, there are many other temporary bat roosts in the African savanna that have yet to be studied.

STRAW-COLOURED FRUIT BAT MIGRATION

Breeding range
• Kasanka National Park
Migratory range
Post-breeding dispersal

MIGRATION PROFILE

Scientific name	*Eidolon helvum*
Migration	To and from seasonal roosts
Journey length	Up to 2,000 km (1,250 miles) each way
Where to watch	Kasanka National Park, Zambia
When to go	November–December

Snow Goose

Each autumn more than six million snow geese head south to escape the big freeze in the tundra. They fly high and fast in family groups that merge into giant squadrons thousands strong, spiraling out of the sky in a maelstrom of beating wings.

Huge blizzards of snow geese are one of the world's foremost wildlife spectacles, and the din created by their constant yelping is equally unforgettable. The largest gatherings of this flourishing species occur each spring and autumn, at traditional migration stopovers on wetlands in the Midwest, USA. Some favoured staging areas host amazing concentrations of geese – for example, 1.2 million were counted at Sand Lake, South Dakota, in April 1991, while 800,000 visited DeSoto National Wildlife Refuge on the Iowa–Nebraska border in November 1995.

LESSER AND GREATER

There are two subspecies of snow geese. Lesser snow geese, by far the most widespread and abundant, have a breeding population of at least five million birds spread throughout the tundra from Wrangel Island, a Russian territory in the Arctic Ocean, east across the top of Canada as far as the western shores of Hudson Bay. Most winter on marshes and low-lying arable fields between northern Mexico and the Mississippi Delta, and in central California. By contrast, greater snow geese have a more easterly distribution and number about a million individuals. The majority breed on Baffin Island and in Greenland, migrating to the USA's middle Atlantic coasts for the winter.

True to their name, snow geese usually have stark white plumage, set off by black wingtips. But scattered within a flock of lesser snow geese there will always be a few dark, slate-grey birds, popularly called "blues". Considered a separate species in the past, these striking birds are now known to be simply a colour variation, or phase, controlled by a single gene.

FAMILY TIES

Snow geese choose a partner when they are two or three years old, and they mate for life – divorces are rare. They nest in large colonies and the females all lay their eggs in tandem. This is an effective strategy for "swamping" their chief predators, Arctic foxes, which are faced with more eggs than they can possibly steal. It also ensures that hatching throughout the colony coincides with the richest time on the tundra, when the lush carpet of grasses and sedges provides plenty of food for the goslings. Within seven weeks, the young geese begin to make their first short, hesitant flights. As the Arctic summer wears on, the geese vacate their nesting areas and congregate in restless flocks along lake shores or river deltas. It is thought that their departure from the far north, which occurs between August and October, may be triggered by shortening day-length, but this theory is as yet unproven.

Each family sticks together on the journey south, with parents and young calling all the time to maintain contact. Many different families will band together into groups of a hundred to a few thousand birds, and often several of these flocks coalesce on the way, eventually forming an impressive wave of as many as 30,000 migrating geese.

Such a concentrated passage of birds can present a serious hazard to airliners, forcing airports such as Winnipeg in Manitoba, Canada to close temporarily while the geese pass.

Facing page Snow geese passing overhead in Vs are an iconic image of migration. Formation flying enables the birds to save energy and share orientation information.
Above A snow goose incubates its eggs in the Arctic tundra, its face stained orange by the high iron content in its summer diet of grasses.

CROWDED FLYWAYS

Snow geese are able to navigate by day or night, coasting at an altitude of around 900 m (3,000 feet) or occasionally much higher; some flocks have been recorded by radar at 6,000 m (20,000 feet). The flocks move in an undulating fashion, with birds ranged at different heights, gently rising and falling as they go, and this peculiar behaviour has earned them the affectionate nickname of "wavies".

Like other migratory waterfowl, the geese bunch together in restricted lanes, or flyways, that follow features of the landscape. Lesser snow geese from the Hudson Bay area, which make up the largest population, use the Mississippi Flyway to commute to and from their Gulf coast wintering grounds. Some may accomplish this flight in only a couple of stages, but in the fall most stop several times for a few days or weeks, until bad weather or food shortages drive them onwards. The spring passage is faster, since the geese have spent weeks feasting on waste corn, laying down substantial reserves to last them during both their migration and the first few weeks in the still-frozen tundra; females may not feed properly until after their 25-day incubation period. In other words, the geese arrive not exhausted but with plenty of gas left in the tank.

MIGRATION PROFILE	
Scientific name	*Chen caerulescens*
Migration	From Arctic breeding areas to temperate winter areas
Journey length	2,000–5,000 km (1,250–3,000 miles) each way
Where to watch	Sand Lake National Wildlife Refuge, South Dakota, USA
When to go	March–April and October–November

FORMATION FLYING

Migrating snow geese organize themselves into U, V, or W shapes. By cruising in the slipstream of the bird in front, geese in formation experience up to 40 per cent less drag, enabling them to lower their heart rate and conserve energy. Every member of the flock has a clear view and flies through "clean" air, avoiding the vortices generated by the flapping wings of the other birds. The leading bird, usually a seasoned adult, has to work hardest, but is replaced at frequent intervals.

Tundra Swan

Tundra swans are the greatest migrants among swans, flying huge distances to and from their remote Arctic breeding grounds. They form lifelong partnerships and unfailingly turn up at the same staging areas and winter haunts year after year.

After the spring thaw, many species of ducks, geese, and swans fly to Arctic ponds to raise young, drawn northwards by the lengthening hours of daylight and the tundra's profusion of plant and invertebrate life. Among the hordes of migratory waterfowl are thousands of tundra swans, whose strident, yodelling calls can often be heard as family groups pass overhead. This evocative sound, a sure sign of the passing seasons in northern latitudes, seems to capture the restless spirit of migration itself and features prominently in Norse and Native American mythology.

Tundra swans occur in two outwardly similar but geographically distinct forms. One subspecies, sometimes known as the whistling swan, breeds from Alaska east to northern Quebec in Canada, while the other breeds in a continuous arc across northernmost Siberia and is named Bewick's swan in honour of the English engraver Thomas Bewick (1753–1828), who produced an early field guide to British birds. Both subspecies in turn are divided into western and eastern populations, which have their own traditional migration routes, staging posts, and wintering areas.

Broadly speaking, North American tundra swans travel south in the autumn to points along either the Pacific or Atlantic seaboard, with the largest numbers heading to the coasts of Washington state and California or to Chesapeake Bay on the East Coast. Their major stopovers are in the northern prairies and lower Great Lakes. Siberian birds show a comparable west–east migratory divide: the western contingent travel via a stopover in the White Sea area to northwest Europe, especially Britain and Ireland; those from further east move to Japan and China.

CREATURES OF HABIT

Tundra swans are remarkably faithful migrants. Most of the marshy coasts, sheltered bays, lakes, and fertile plains to which they flock each winter and on migration have hosted the species throughout recorded history, and probably long before that, in some cases possibly as far back as the last ice age 15,000 years ago. This extraordinary demonstration of philopatry – the tendency of a migratory animal to return to a particular location – has several advantages for the swans. Above all, familiarity with an area means they know the best feeding places and the safest roost sites.

Above Every tundra swan hatches with a migratory instinct and continually enhances its orientation ability with cues picked up during its seasonal journeys.

In tundra swans, like other swans and geese, migratory patterns are passed from generation to generation. Tundra swans are long-lived birds (the oldest on record was a 36-year-old swan from Siberia), they take two or three years to reach breeding age, and the young depend on their parents for up to 10 months after hatching in June or early July. Each family migrates south together so that the cygnets can learn the route. The family remains as a unit all winter, and, still united, returns north again in spring. Finally, having arrived at the nesting grounds, the immature birds are at last driven away by their parents. Usually they move to communal feeding and moulting areas, gathering in flocks made up of other non-breeding birds, but in late summer it is not uncommon for immature birds to rejoin their parents with a new brood of young in tow. The resulting "superfamily" migrates as a group. Some of these extended families contain young from three different seasons.

HIGH FLIERS

Tundra swans are purposeful, powerful migrants that travel by both day and night. It has been calculated that their bodies carry enough

Above Tundra swans migrate as a family, with adults guiding their young on their first trips south in autumn and north again in spring.

TUNDRA SWAN MIGRATION

Whistling swan breeding range
Whistling swan winter range
Bewick's swan breeding range
Bewick's swan winter range
↔ Migration routes

fat for a non-stop flight of 1,500 km (900 miles), equivalent to 24 hours' flying time at an average speed of 60 kph (37 mph). They fly at relatively low altitudes of less than 450 m (1,500 feet) to avoid an exhausting climb, forming tight-knit formations to achieve maximum aerodynamic efficiency. Scientists who compared radio-tracking data with meteorological records found that the swans have to wait patiently for the right conditions – they are effectively grounded until a suitable tailwind enables them to set off together on the next leg of their impressive journey.

MIGRATION PROFILE

Scientific name	*Cygnus columbianus*
Migration	From Arctic breeding areas to temperate winter areas
Journey length	2,500–5,000 km (1,600–3,000 miles) each way
Where to watch	Southern Great Lakes, Canada/USA; Freezeout Lake, Montana, USA
When to go	March–April and October–November

UNIQUE PATTERNS

In the mid-1960s, during a pioneering study of Bewick's swans in England, conservationist Peter Scott (1909–1989) realized that the bill markings on each swan were slightly different. The arrangement of black and yellow was unique, enabling Scott to identify individual birds in his study area at a glance. Over several decades researchers from the Wildlife and Wetlands Trust (WWT), founded by Scott, have expanded this database by compiling every swan's life story, including its typical autumn arrival and spring departure dates, behaviour, diet, choice of partner, and breeding success (indicated by the number of young).

Above The unique bill patterns of Bewick's swans enable researchers to track the comings and goings of named individuals.

Wave Riders

Seventeen of the world's 21 albatross species live in the Southern Hemisphere. The Southern Ocean is an inhospitable realm of relentless gale-force winds and towering waves, and the albatrosses turn this stormy environment to their advantage, by capturing the ocean's energy to soar for hours on end.

Sailors have long used the names "Roaring Forties" and "Furious Fifties" for the fierce seas below Patagonia, the Cape of Good Hope, and southern Australia. There are few landmasses to slow the circulation of air, and as a result this is the windiest place on Earth. During the heyday of sail, clippers would head south to make the fastest possible circumnavigation of the globe, and for birds too, the incessant wind at higher latitudes provides the means to travel immense distances with extreme speed. None are better suited to wind-assisted crossings than the albatrosses, which roam thousands of miles to search for schools of squid, their main prey.

Albatrosses have the longest wings of any birds – the wandering albatross, the family's largest species, has an amazing 3.5 m (11 feet) wingspan. Their wing width is reduced to the bare minimum, as in a glider, to minimize drag for aerial efficiency. Masterful fliers, these elegant birds can maintain average speeds of 55 kph (34 mph) for up to 12 hours at a stretch. To conserve energy, they skim across the ocean surface to seek the uplift created by air forced up over the waves. Riding one updraft after another, they coast along with scarcely a wing beat. Yet albatrosses are poorly designed for continuous flapping flight – on rare occasions when the wind drops, they must rest on the water. This means that Southern Hemisphere species seldom manage to cross the equator, due to a belt of becalmed air called the Doldrums.

Facing page Albatrosses are the ultimate long-distance nomads, adapted to glide fast at a low energetic cost. A special tendon keeps their outstretched wings locked in place.

Inset left Having made landfall, a wandering albatross performs its greeting display. Like the rest of the family, this species breeds on remote oceanic islands.

Inset left, below This photograph of a soaring grey-headed albatross demonstrates the phenomenal wingspan for which these seabirds are famous. Some species routinely circle the Earth to track down food.

Below One of the most efficient flight techniques over water is dynamic soaring. An albatross wheels into the wind to gain height (1), changes tack (2) to glide downwind for as long as it can (3), and then, having descended almost to sea level, it neatly catches an updraft from a wave to repeat the manoeuvre (4).

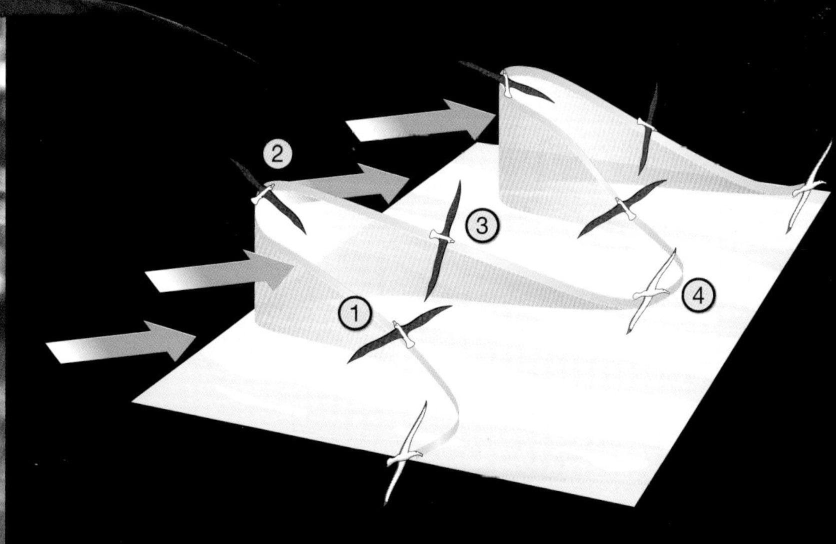

Short-tailed Shearwater

Soaring just above the waves, this tireless ocean nomad flies immense distances during its lifetime, skirting almost the entire North Pacific to complete its annual migratory circuit. Its six-month journey ranks among the greatest of all bird migrations.

The short-tailed shearwater is exceptionally efficient at gliding across the sea: its long, narrow wings maintain high speeds for hours on end; and each wing has curved surfaces to produce an aerofoil shape that generates lift. It is also a powerful swimmer, using its wings for propulsion underwater, but usually it feeds at the surface on krill and other crustaceans. However, the shearwater moves clumsily on land, so is vulnerable to predators. As a result, it touches down only in darkness and nests in burrows for protection.

RETURN TO LAND

All of the world's adult short-tailed shearwaters, an estimated 23 million individuals, converge on the waters off southern Australia to breed. They nest in crowded colonies on exposed islands and headlands, and for about six months of the year are the region's most abundant seabird. Most of the colonies lie in Bass Strait, the windy channel of shallow water that separates the Australian mainland from Tasmania. At Fisher Island, in the east of the Strait, ornithologists have systematically monitored every bird nesting in a small, 100–200 pair colony for more than half a century. The shearwaters are extremely faithful to their breeding site: over 40 per cent of the young birds hatched here later return to breed, often to the same small part of the colony.

Groups of shearwaters start arriving offshore in mid-September and the colony is already full by the month's end. The peak arrival varies little from year to year – a remarkable feat of synchrony for

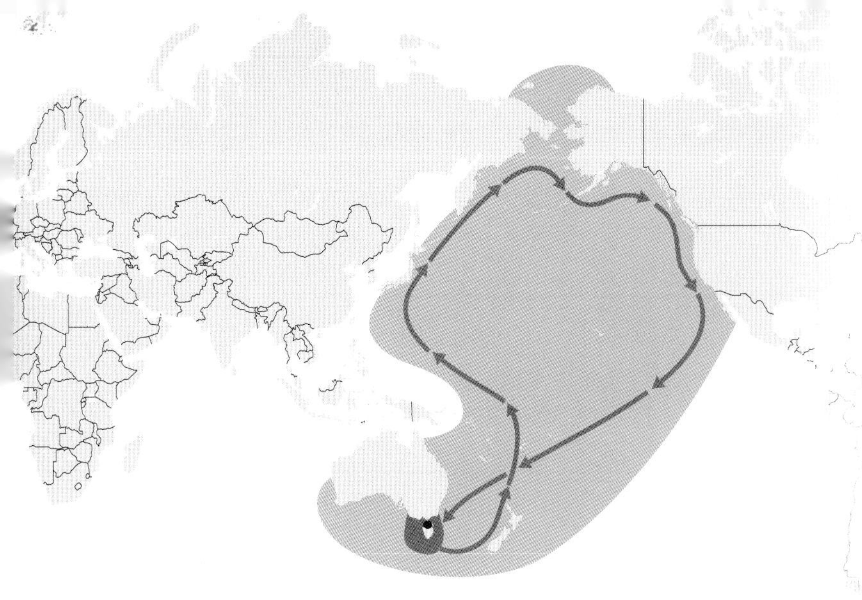

SHORT-TAILED SHEARWATER MIGRATION

Breeding range · Migratory range

● Bass Strait → Migration route

Facing page Short-tailed shearwaters often assemble in rafts thousands strong near their breeding colonies or while they are feeding at sea. From afar, these impressive gatherings resemble dense oil slicks.

MIGRATION PROFILE

Scientific name	*Puffinus tenuirostris*
Migration	Clockwise loop around Pacific
Journey length	11,250–16,500 km (7,000–10,250 miles)
Where to watch	Bass Strait, between Australia and Tasmania
When to go	November–December

such a wide-ranging migratory seabird. In the last week of November each mated pair lays a single egg, which hatches after a 53-day incubation. Both parents go fishing for their chick, travelling south to within 80 km (50 miles) of the Antarctic ice-edge on trips that last up to 17 days, and the youngster grows rapidly on its diet of semi-digested fish and krill.

After 10 weeks of heavy feeding, the corpulent chick may weigh twice as much as its parents. Given that a large shearwater colony contains hundreds of thousands of equally bloated, oil-rich birds, it is a valuable source of protein for local people. These shearwater chicks, known as mutton-birds, have long provided a sustainable harvest for the residents of Tasmania, who continue to collect about 200,000 every year.

OCEAN ODYSSEY

In early April the adult shearwaters desert their colonies to begin the first leg of their trans-equatorial migration, which takes them through the Tasman Sea to the west of New Zealand. Their abandoned young grow ever hungrier, slimming down to their optimum flight weight while simultaneously developing their first set of flight feathers. Hunger drives them to explore outside their burrows and exercise their wings, until two or three weeks after the parents' departure, they too fly out to sea. Learning evidently plays no part in the migratory make-up of young shearwaters: they navigate instinctively, without the assistance of experienced birds.

The shearwaters embark on an extraordinary migration around the Pacific Ocean, taking six or seven months to trace an enormous circle. In the first stretch, they travel past the Philippines and move through the waters off East Asia, Japan, and Russia's Kamchatka Peninsula, then they cross to the western seaboard of North America and follow it south towards California, before finally traversing the mid-Pacific on a southwesterly course to return to Australia. Each stage of this looping migration route takes advantage of prevailing winds and seasonally changing food supplies. Many shearwaters spend the northern summer in the icy, productive waters of the shallow Bering Sea, where the huge shoals of capelin and Arctic krill are their chief prey. Some birds have been seen even further north, among the ice floes of the Arctic Ocean.

Since the average lifespan of short-tailed shearwaters is 15–19 years, and they typically migrate 14,500 km (9,000 miles) in a year, they are likely to have traveled roughly 246,500 km (153,000 miles) by the time they die, not including feeding trips from their breeding colonies. The oldest individual on record survived for 38 years, so could have flown over twice as far.

TUBENOSES

Shearwaters, petrels, and albatrosses are known collectively as "tubenoses" for their specialized bill design. In most other birds the nostrils are indistinct, appearing as a pair of small openings near the base of the bill, whereas these maritime wanderers have large, tubular, external nostrils that often form a prominent ridge. The presence of such well-developed nasal tubes is a clear sign that tubenoses – unlike most of the world's land birds – have a superb sense of smell. They are especially sensitive to the odour of animal fats, and can detect food floating on the water from great distances; when a fishing boat throws offal overboard or when a dead whale floats rotting in the water, flocks of these seabirds will suddenly materialize. Tubenoses recognize their nest-sites by smell, so do they also use it for long-distance orientation? Perhaps, but most ornithologists doubt that their olfactory ability is this effective.

Below Shearwaters have a highly developed sense of smell and can pick up the faint whiff of distant food as they skim low over the sea.

Manx Shearwater

A ten-week-old Manx shearwater may take little more than a fortnight to fly across the Atlantic to reach its wintering area. It navigates through empty stretches of ocean purely by instinct, and later will return to breed on the same island on which it hatched.

Manx shearwaters live for up to 50 years, and, apart from brief shore visits during the nesting season, spend their whole adult life at sea. These lithe, long-winged, fish-eating birds are perfectly in tune with the marine environment. They can even drink seawater, because of their specialized salt glands that process the unwanted salt and expel it through their nostrils. More impressive still is their capacity to navigate rapidly and accurately across immense expanses of unchanging ocean.

Recoveries of banded Manx shearwaters demonstrate that the species can fly from its breeding colonies in the northeast Atlantic to its winter quarters in the tropical seas of eastern South America within two to three weeks, provided there is a favourable tailwind. The chicks are hard-wired with this extraordinary ability. One young shearwater, ringed in Wales as a fledgling, was picked up in the waters off southern Brazil 16 days later, an estimated three days after it had died. It had therefore travelled 9,600 km (6,000 miles) at a rate of 740 km (460 miles) per day. For it to make such quick progress, the bird could not have wasted much time getting its bearings; some kind of mental route-map must have helped it to identify the precise direction it needed to head in.

STRANGE CHORUS

Three small rocky islands support 70 per cent of the global breeding population of Manx shearwaters. Rum, which lies off northwest Scotland, is used by 100,000 nesting pairs, while the adjacent islands of Skomer and Skokholm on the coast of west Wales hold

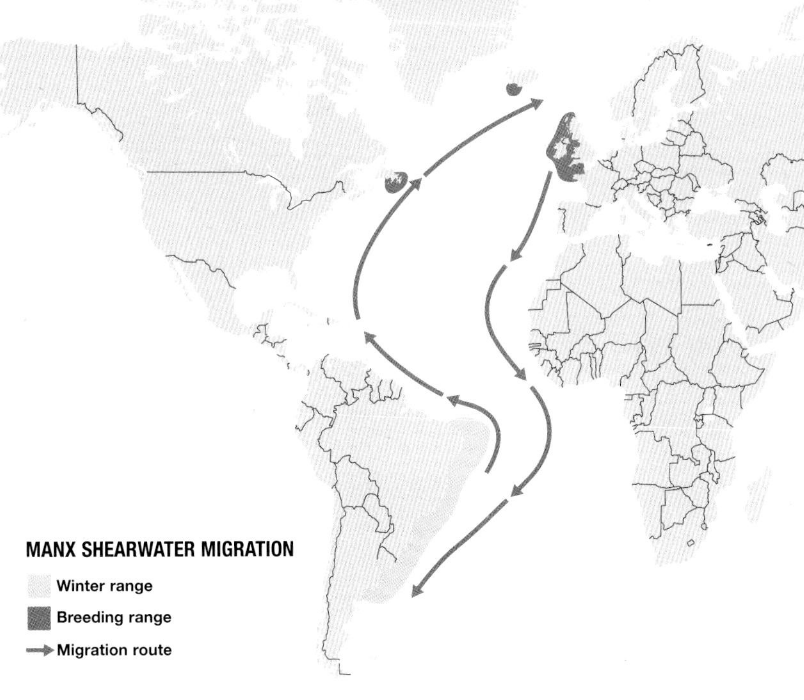

135,000 pairs between them. At night when the shearwaters return to their colonies from fishing trips the darkened skies echo with a cacophany of harsh cries. This noise lead 11th-century Vikings to believe that Rum was inhabited by trolls.

Manx shearwaters of breeding age – those at least six years old – reoccupy their colonies in March. The males clean out their burrows, then meet up with their lifelong partners (a minority, usually birds that failed to raise young the previous year, will "divorce" and choose a different mate). By mid-May every female has produced a large egg, weighing about 15 per cent of her body weight, and each pair initiates a shift system that lasts throughout the seven to eight week incubation, with one bird staying in the burrow to incubate while the other flies off for several days to feed. This routine continues after the chick hatches, but as it gets older both adults go fishing daily to satisfy its huge appetite.

FAT CHICKS

In many seabird species, the chicks gain weight very fast, and young Manx shearwaters are no exception: at their peak, they weigh 20 per cent more than their parents. Their fat reserves enable them to migrate immediately after leaving their burrows in August or September and tide them over until they finally arrive at their destination. They can finish the long journey without feeding en route – a big advantage for inexperienced young birds that would have difficulty finding enough to eat in mid-ocean, where food is always scarce.

Juvenile Manx shearwaters spend their first year at sea, but from their second summer onwards start to visit the colony where they hatched. They show very high levels of natal philopatry – that is, the tendency of an individual bird to return to its natal area. In the 1950s, researchers conducted a famous "displacement experiment" in which shearwater chicks from Wales were flown by airplane to Boston, Massachusetts, USA then released and tracked. One of the birds promptly returned to Skokholm, 4,900 km (3,050 miles) across the Atlantic, in under 13 days.

More recently, experiments on Cory's shearwaters have proved that they can find their way home even with magnets attached to their head and wings, to disrupt their perception of the Earth's magnetic field. It seems that members of this seabird family possess a site-dependent homing system that does not rely on magnetic clues.

Facing page The shearwaters ditch into the sea to rest and socialize, especially at dusk when waiting to come ashore, but their migrations are speedy affairs carried out with few breaks.

MIGRATION PROFILE

Scientific name	*Puffinus puffinus*
Migration	Circuit around North Atlantic
Journey length	8,500–13,000 km (5,250–8,000 miles) each way
Where to watch	Skomer Island, Wales, UK
When to go	May–June

Above Flocks of Manx shearwaters produce a distinctive flickering effect as they fly due to the alternation of the birds' white underparts and black upperparts.

RAT THREAT

On land, Manx shearwaters are a tempting target for large predatory seabirds such as great black-backed gulls and great skuas. By coming ashore after nightfall, most shearwaters avoid these diurnal predators. However, this strategy is no defence against nocturnal mammalian hunters, particularly egg-scavenging rats. For this reason the birds thrive only on inaccessible, rat-free islands.

Above They are masters in the air, but Manx shearwaters are a rather pathetic sight on land, usually landing in a crumpled heap.

European White Stork

According to legend white storks are the bearers of human babies, and they have been a symbol of fertility throughout Europe for thousands of years. Many festivals honour the reappearance of these majestic birds each spring, in huge flocks travelling out of Africa.

White storks are a familiar sight across much of southern, central, and eastern Europe in spring and summer. They can often be seen soaring high overhead on massive wings or striding through fields and marshes in search of prey, despatching victims with a strike of their blood-red, dagger-shaped bill. Great opportunists, storks have benefitted hugely from their ancient association with humans and long ago gave up nesting in trees, preferring the safety and convenience of rooftops and towers. Today, they use all manner of artificial structures as nest-sites, from chimneys to road signs, pylons, and radio transmitter masts, but church spires remain a favourite.

CLOSE TO PEOPLE

In parts of Spain and eastern Europe almost every cathedral and chapel has its own pair of nesting storks, causing one commentator to remark that the Catholic Church may be the most successful stork conservation organization in history. Local communities in these areas have tremendous affection for "their" storks, which arrive on cue in about the first week of April and leave promptly at the end of August. Accustomed to this protection, the storks often conduct their lengthy bill-rattling courtship and raise their young in full view of the street below, even in busy town centres.

Ringing studies have proved that, while there is some truth in the traditional belief that storks mate for life and meet up again at the same nest each spring, an individual bird is more loyal to its nest-site than its partner. Over many seasons, a nest will therefore be used by a variety of males and females from successive generations of storks. Old nests develop into giant mounds of sticks, and some eventually reach a weight of about half a ton.

EAST AND WEST

Most European white storks move to sub-Saharan Africa for the winter. In common with other migratory storks, cranes, and pelicans, and most of the world's raptors, they travel primarily by soaring. Their broad wings are adapted for riding thermals, enabling the storks to spiral upwards effortlessly, despite their considerable weight. Soaring is so efficient that storks have no need to indulge in pre-migratory fattening – they expend little more energy on migration than in their normal daily activity. Thermals do not form over water, however, so the trade-off for relying on soaring flight is that they cannot accomplish long sea crossings.

The storks are restricted to one of two routes to and from Africa. Those from western areas funnel south through Spain to cross the Mediterranean at the Strait of Gibraltar, a short distance of only 8 km (13 miles). Meanwhile, eastern breeders, which outnumber the western contingent by a factor of 10, head southeast to cross from European to Asian Turkey at the Strait of Bosporus, then skirt the eastern shores of the Mediterranean to enter Africa through Israel and Sinai.

Storks fly in the heat of the day, when thermals are strongest, and bunch together in spectacular numbers along these narrow fronts – one wave of storks passing over the Turkish city of Istanbul on the Bosporus contained 11,000 birds. In all, about 350,000 white storks move through the Bosporus during the species' month-long autumn exodus, with 35,000 crossing at the Strait of Gibraltar. Once in Africa, the storks still have several thousand miles to go before they reach their winter quarters, but the hot climate creates plenty of thermals, so they can make rapid headway.

STORK RESEARCH

Since the mid-1990s, researchers studying white storks have had access to solar-powered platform transmitter terminals (PTTs), which offer a far superior lifespan to battery-operated devices. The resulting data have shown that the storks are largely nomadic during the non-breeding season, and may winter in different parts of Africa from one year to the next. One female stork fitted with a long-life PTT has been tracked continuously for over 10 years, in which time she has made six return-trips between Africa and her German nesting area. This is currently the longest period for which any individual animal has been followed using satellite-based radio-tracking.

Facing page White storks travel en masse, bunching together at migratory bottlenecks around the shores of the Mediterranean Sea. **Inset left** Many local communities put up rooftop platforms to encourage storks to nest.

MIGRATION PROFILE

Scientific name	*Ciconia ciconia*
Migration	From European breeding grounds to African wintering areas
Journey length	2,000–10,500 km (1,250–6,500 miles) each way
Where to watch	Strait of Bosporus, Turkey
When to go	Mid-August–mid-September

STUDYING A POPULATION

The UK osprey population is small in global terms, with about 200 breeding pairs in 2007, but almost every nest has been monitored since 1954. During half a century of research, more than 1,250 British ospreys have been banded at their nest-sites, and some fitted with numbered, colour-coded wing tags that can be read through binoculars. Records of marked birds have built into a fascinating picture of the species' migration through France and Spain to West Africa, including the different journeys made by males, females, and non-breeding juveniles.

Today, the equipment used in satellite telemetry is capable of sending hourly data on an osprey's location, direction, speed, and altitude, and the GPS-fix is accurate enough to pinpoint the seashore or body of fresh water on which it is fishing. Osprey migration websites carry live updates of individual satellite-tagged birds, generating valuable publicity for conservation.

Left Osprey chicks are measured, weighed, and banded before being returned to the nest.

Osprey

A supremely graceful flier, the osprey patrols stretches of water with relaxed, shallow wingbeats to hunt for fish. Unlike other birds of prey, it often makes long sea crossings and tends to spread out when travelling, rather than flocking to narrow migration corridors.

For taxonomists, the osprey is something of a puzzle. This unusual bird of prey, a fisher *par excellence* that also goes by the colloquial name of fish hawk, has no close relatives, and as a result is placed in a genus of its own. Its diet is up to 99 per cent live fish, which it catches after a dramatic swoop from a height of 10–30 m (30–100 feet), plunging feet-first into the water with a huge splash. A reversible fourth toe and spiny-soled feet help it to grip its twisting, slippery prey while it flies to a nearby perch to consume its meal.

MIGRATION PROFILE

Scientific name	*Pandion haliaetus*
Migration	Northern populations move south in winter
Journey length	4,000–10,000 km (2,500–6,250 miles) each way
Where to watch	Delmarva Peninsula, Virginia, USA; Falsterbo Peninsula, Sweden
When to go	September (Delmarva); August–September (Falsterbo)

Facing page Migrating ospreys stop on the way to fish at productive hunting grounds. **Right** Mated pairs of ospreys return to the same nest year after year, but migrate separately. Each pair can raise up to four young.

FLEEING THE COLD

The osprey is one of the world's most widespread birds, being found on every continent except Antarctica , including on many oceanic islands. It has resident, non-migratory populations in the tropics and subtropics, for example in southern Florida and the Gulf coast states, California, parts of the Middle East, and Australia, but in the rest of its enormous range is usually a migrant. The key factor driving its migrations is wintry weather, because cold conditions force fish into deeper water where they are out of reach. For an osprey, an iced-over lake or river spells certain death.

All the Canadian and American ospreys that breed in latitudes above 30–32° N vacate their nesting areas in the fall. They move south to California, coasts in the Caribbean, and the northern half of South America, and to the rivers and lagoons of the Orinoco and Amazon drainage basins. In Europe, the dividing line between migratory and non-migratory osprey populations lies further north, at about 38–40° N. The north European birds winter in Africa south of the Sahara Desert, especially along Atlantic coasts near the equator. Ospreys from Russia and Japan head mainly to the shores of Arabia, the Indian subcontinent, and Southeast Asia, and frequent mangrove swamps in large numbers.

SLOW PROGRESS

Ospreys travel by day, and move at a slower pace than most migratory hawks and eagles of comparable size. They pause at favourite feeding sites along the way – often for several days or a week – and it seems that different individuals use different stop-off areas, with each bird visiting its preferred estuaries, lakes, or marshes year after year. There have been instances of a ringed osprey returning to exactly the same perch during the same one- or two-week period of its annual migration, much to the delight of local ornithologists.

Migrating ospreys travel alone and follow their own route – another factor that distinguishes this species from other migratory raptors, which tend to form flocks. The ospreys typically fan out across a wide area, known as a broad front. For this reason, day counts of ospreys passing over hawkwatch points tend to be much lower than for other raptors. One place that does witness impressive numbers of migrating ospreys is the Delmarva Peninsula, on the eastern side of Chesapeake Bay in coastal Virginia, where more than a hundred can be seen on a good day in September. Falsterbo, a peninsula at the southern tip of Sweden, is also a reliable location for watching movements of ospreys, in late August and early September.

Most raptors and other large soaring birds avoid lengthy sea crossings at all costs, since the thermals and upcurrents they need do not form over water. Ospreys, however, are powerful fliers that readily fly over water barriers. Scottish and Scandinavian ospreys flying south in the autumn often take a short-cut across the Bay of Biscay to reach Spain, while some American ospreys migrate across the Gulf of Mexico. There are even records of ospreys up to 100 km (60 miles) off the USA's Atlantic coast.

Swainson's Hawk

Swainson's hawk travels as far as any other North American bird of prey and devotes a third of its life to migrating. Every year most of its population flies from the Great Plains to the pampas of Argentina and back, in great flocks containing thousands of soaring birds.

Named after the English naturalist William Swainson (1789–1855), this hawk is a slim raptor with a long tail and long, pointed wings that it holds in a distinctive shallow "V" when soaring. It is a characteristic bird of open country, especially grassland, and its breeding heartland is the prairies of western North America. Although it ranges north as far as Alaskan tundra, and south to northern parts of Mexico, the species is commonest on the Great Plains, where from April to September it is a familiar sight perched on telephone poles, fence posts, and dead trees.

Field observations and data from ringing studies suggest that mated pairs of Swainson's hawks are faithful to both each other and their nest-site. The partners return independently to the same spot each spring despite having travelled up to 29,000 km (18,000 miles) during seven months apart, reaffirming their pair-bond with a series of noisy, circling and diving courtship displays. Such a degree of breeding philopatry, or nest-site fidelity, is remarkable in an extreme long-range migrant.

SWITCH IN DIET

Throughout the nesting season, the hawks rely on their skill as expert rodent catchers, with first the male and then both parents bringing a regular supply of ground squirrels, gophers, mice, and baby rabbits for the young. But at other times, these raptors are specialist insect predators. In their wintering grounds on the grassy plains of Paraguay and northern Argentina, the hawks prey mostly on grasshoppers and migratory dragonflies, and roam widely to track the insect hordes.

A precipitous decline in Swainson's hawks in the 1980s and 1990s was traced back to problems in their wintering range: heavy

Above Migrating hawks avoid crossing open water, so all of the birds travelling to and from South America have to pass over the narrow strip of land at the Panama Canal.

This page This impressive flock of Swainson's hawks was photographed at Panama's Ancon Hill during the birds' autumn migration.

MIGRATION PROFILE

Scientific name	*Buteo swainsonii*
Migration	Most move to Argentina for winter
Journey length	6,000–14,500 km (3,750–9,000 miles) each way
Where to watch	Hazel Bazemore County Park, Corpus Christi, Texas; Ancon Hill, Panama City, Panama
When to go	Late September–October (Texas); October–early November (Panama)

SWAINSON'S HAWK MIGRATION

Breeding range

Main winter range

Secondary winter range

Breeding and wintering

Migration route

use of organophosphate pesticides reduced the raptors' primary food source and many birds died of poisoning. The species has since begun a slow recovery, due partly to conservation measures in Argentina, but also perhaps to changes in its migration patterns. Some hawks seem to be wintering in Brazil north of the main pampas region, while small, possibly growing, numbers are now spending the winter in southern Florida, California, and Central America. Whether this shift in behaviour is a direct consequence of agricultural intensification in the pampas, or the result of another process, such as climate change, remains uncertain.

THERMAL SOARING

During the four months of the year that they spend travelling, Swainson's hawks are highly gregarious. Their enormous flocks, known as "kettles", can resemble dense swarms of insects from afar, and are easily observed from traditional hawkwatch sites, such as Hazel Bazemore County Park in Texas. Twice each year, in March–April and October–November, the entire South American wintering population of Swainson's hawks – about a million birds – passes over Ancon Hill near Panama City.

These dramatic mass movements happen when spells of warm, dry weather create thermals, enabling the hawks to soar with minimum effort. In Panama, the hawk flocks take advantage of long sheets of cloud produced by bands of rising air, and glide through the cloud base for tens of miles with barely a wingbeat. The power saving that soaring and gliding offers over flapping may be as great as 95–97 per cent. This energy conservation is important, because Swainson's hawks flying to Argentina fast for up to 60 days (there are few records of them feeding en route and no droppings have been found at their communal roosts). The simple fact that the birds can arrive without once feeding proves the low energetic cost of their long-haul journey.

Above After years of decline, Swainson's hawks have staged a welcome recovery in the prairies of North America. They are often seen perched on artificial structures such as fences, which provide an ideal look-out for prey.

MIGRATORY BOTTLENECKS

Each year in the autumn, half a million migrating birds of prey stream through the Mesoamerican corridor, which stretches from southern Mexico to Panama. This strip of land narrows dramatically in the south, forcing the migrant raptors to bunch together in ever-greater concentrations. The land bridge is narrowest at the Isthmus of Panama and in the Panama Canal area is a mere 50 km (30 miles) wide. Four species dominate the raptor flocks here: the broad-winged hawk; turkey vulture; Swainson's hawk; and Mississippi kite (in that order). They are joined by smaller numbers of up to 24 other North American raptor species. Spectacular flocks of migrating raptors also occur at Whitefish Point in Michigan, where a thin strip of land divides lakes Michigan, Huron, and Superior; the Strait of Gibraltar, which forms the gateway between Spain and North Africa; and the Strait of Bosporus, which separates European and Asian Turkey.

Land of Flood and Fire

Violent tropical storms collide with northern Australia each summer, unleashing a deluge that creates mile after mile of lagoon and swamp. Countless thousands of migratory birds flock to the short-lived wetlands to feed and breed.

No other large land mass burns as often as Australia, where spectacular bush fires have a profound impact on the vegetation, but the northernmost quarter lies within a belt of low pressure called the Intertropical Convergence Zone (ITCZ), which brings four months of monsoon rains. From December to April, the Australian north coast is battered repeatedly by storms that cause ephemeral rivers to flow once more and spill over their banks. Woodlands are inundated, trees burst into flower, and low-lying floodplains become inland seas carpeted with waterlilies, hyacinths, and wild rice plants.

This transformation signals an avian invasion on a grand scale. One of the best places to witness the migration is the coastal portion of the Northern Territory, known colloquially as the "Top End". At the heart of this wilderness, about 250 km (150 miles) east of the regional capital Darwin, is Kakadu National Park. The park's groves of blossoming paperbark trees soon attract nomadic flocks of nectar-feeders, such as red-collared lorikeets. However, the most significant migrants, in terms of numbers, are water birds. Several dozen species congregate on lush wetlands in the park, including magpie geese, black swans, wandering whistling-ducks, and Australian pelicans. Some come from the surrounding area, while others fly thousands of miles across the continent. They are joined by visitors from the Northern Hemisphere – waders such as curlews and sandpipers that breed in the Arctic tundra. The long-haul champions are bar-tailed godwits, which make a journey from their nesting grounds in the bogs of northwest Alaska, up to 14,500 km (9,000 miles) away.

Left The rains turn low-lying areas of Kakadu National Park into a vast transient wetland, attracting a host of migratory water birds.

Below left Magpie geese graze wild rice and other aquatic plants and, like many of the other visiting birds, waste no time in raising their young before the floodwaters recede.

Below Australia's "Top End" feels the full force of the Indian Ocean monsoon and so has some of the highest rainfall, with a yearly average of 800 mm (31 inches) in the south of the region and 1,200 mm (47 inches) in the north. The centre of the continent is arid desert.

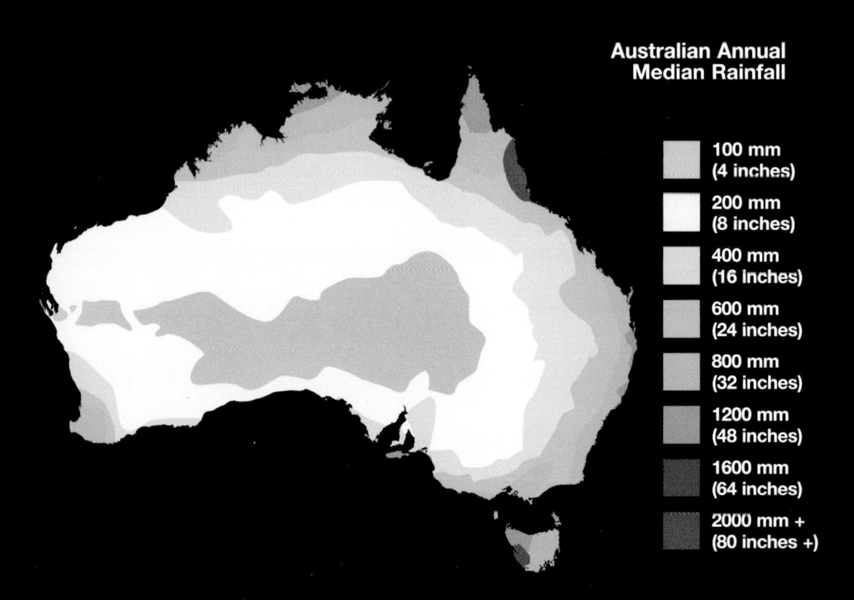

Whooping Crane

Pushed to the brink of extinction by hunting and loss of habitat, North American whooping cranes remain at risk today. The struggle to rescue them illustrates the difficulty of conserving migratory animals that travel long distances and need protection every step of the way.

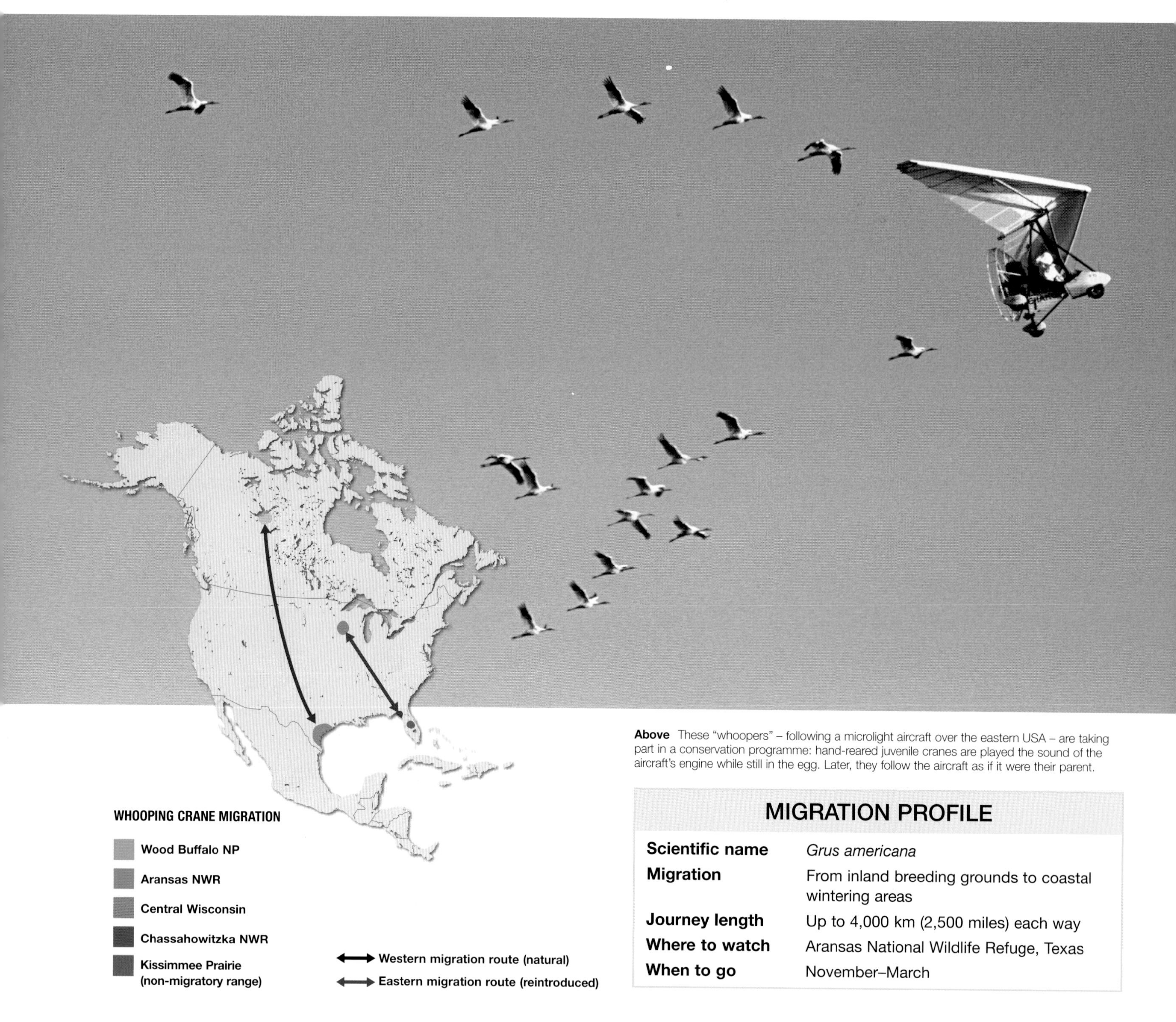

Above These "whoopers" – following a microlight aircraft over the eastern USA – are taking part in a conservation programme: hand-reared juvenile cranes are played the sound of the aircraft's engine while still in the egg. Later, they follow the aircraft as if it were their parent.

MIGRATION PROFILE

Scientific name	*Grus americana*
Migration	From inland breeding grounds to coastal wintering areas
Journey length	Up to 4,000 km (2,500 miles) each way
Where to watch	Aransas National Wildlife Refuge, Texas
When to go	November–March

Whooping cranes are an ancient species – their fossilized remains date back several million years – and the largest birds native to North America, at about 1.5 m (5 feet) tall, but unfortunately have become better known for their great rarity. These stately, snow-white birds serve as a tragic symbol of the environmental damage inflicted on the American wilderness over the last two centuries.

Named after the eerie, far-carrying sounds uttered by mated pairs, whooping cranes were once a relatively common sight throughout the Canadian prairie provinces and plains of the Midwest, and along the USA's Atlantic coast. They began a steep population decline in the 1800s as a result of the drainage of their marshy feeding and nesting areas to create farmland, unregulated shooting by waterfowlers, and trophy collecting for taxidermy. By 1941, only 15 individuals were left. All of the "whoopers" alive today – wild and captive – are descended from this tiny group.

LAST SURVIVORS

The handful of cranes that held on into the 1940s were the remnant of a much larger migratory population spread across the continent. For a long time, the survivors' precise nesting area was a mystery, then in 1954 it was tracked down to a remote region of bogs and rivers in Wood Buffalo National Park, in Alberta and Northwest Territories. Conservationists already knew that this relict population wintered on the salt-marshes of the coastal bend of Texas. Many more details of these cranes' migration have since emerged, pieced together from patient field work, ringing data, and (in recent years) satellite telemetry.

One of the most remarkable aspects to whooping crane migration is how consistent it is. The cranes are evidently very accurate navigators, keeping to a well-defined, narrow path that juveniles learn from their parents, and which hardly varies from year to year. They proceed along this flyway in a series of 300–500-km (185–300-mile) stages, and travel by day, as a single pair or family party, occasionally in the company of sandhill cranes.

REBUILDING A MIGRATORY SPECIES

Intense conservation efforts over 50 years have produced an impressive recovery in the number of whooping cranes migrating between Canada and Texas. The best place to see this thriving flock is Aransas National Wildlife Refuge (NWR) in southern Texas, where a total of 266 individuals wintered in 2007–2008. However, attempts to create new whooper populations elsewhere, as an insurance policy against disasters striking the natural flock, have met with mixed success.

In the 1970s and 1980s, eggs collected from whooper nests in Wood Buffalo National Park were translocated to the nests of sandhill cranes in Idaho. The plan was for the sandhill cranes to raise the young whoopers and teach them the traditional migration route of sandhill cranes to their wintering area at Bosque del Apache National Wildlife Refuge, near the Rio Grande in New Mexico. But the whoopers, imprinted on their foster parents, regarded themselves as sandhill cranes and failed to develop normal pair-bonds with others of their kind. Few of the fostered whoopers survived and the project was abandoned.

After this setback, a captive-breeding programme was established to hand-rear young whooping cranes for reintroduction to the wild.

TRADITIONAL STOPOVERS

Generations of whooping cranes have used the same staging posts on migration. Birds leave Wood Buffalo National Park in late September or early October, taking only a few days to reach their first major stopover – the wheat belt of southern Saskatchewan. Here, they rest and refuel for several days or weeks, building their energy reserves. Usually they leave Saskatchewan by the end of October, to head south over the Great Plains to their second important stopover, the central Platte River in Nebraska. From here, the whoopers push on to the Texas coast, generally arriving by early December.

Above Whooping cranes can be distinguished from sandhill cranes by their larger size and white plumage.

Throughout the 1990s, cranes were released at Kissimmee Prairie in Florida, building into a small population of 54 birds by 2006. The Florida flock, in common with a non-migratory population that used to exist in the swamps of Louisiana, is present all year round, but in 1999 a coalition of government agencies and NGOs decided to go one better by launching an ambitious scheme to set up a self-sustaining migratory flock of whooping cranes in the eastern USA, where the species died out in the 19th century.

Recreating a migration route that has been completely lost, using hand-reared chicks with no knowledge of the historical journeys undertaken by their ancestors, poses an immense challenge. The solution has been to train juvenile birds to follow microlight aircraft, in order that they can be guided along a safe route and shown the "correct" stopovers to use. Having made the southbound trip, the cranes carry out the return leg and all future migrations unaided. By winter 2007–2008, there were 76 whoopers commuting annually between Wisconsin and Florida. Groups of each year's new cranes cruising within feet of their "parent" microlight have become one of the strangest spectacles in North American skies.

Knot

The knot spends seven months of every year on the move. It flies to some of the northernmost solid ground on Earth to breed, then divides the rest of its time between a network of estuaries and coastal lagoons thousands of miles to the south.

TIME OF PLENTY

Above The horseshoe crab egg-laying season leaves behind a protein-rich banquet for migratory flocks of birds.

Flocks of knots moving up eastern coasts of the USA in spring time their northward advance to hit Delaware Bay, New Jersey, at the end of May, when millions of horseshoe crabs come ashore to spawn. For a few weeks the beaches teem with knots, all frantically probing the mud and sand to find the nutritious crab eggs. During this feeding frenzy, the birds boost their body mass by over half – enough to fuel the last leg of their migration to the Canadian Arctic. Delaware Bay hosted peak counts of up to 95,000 knots in the late 1980s, but numbers fell dramatically over the following decade due to overfishing of the crabs, which depleted the birds' food supply. In 2006, neighbouring states imposed a moratorium on the horseshoe crab harvest to allow crab stocks to recover.

Each summer there is an explosion of insect life in the tundra belt of the High Arctic, which creates a window of opportunity for long-haul migrants. In June and July this vast wilderness of shallow pools and bogs is bathed in perpetual daylight, enabling the birds to feed around the clock and raise a family much faster. Furthermore, there is no shortage of nest-sites this far north and fewer predators.

Many waders migrate to the tundra to produce young, but the knot, a medium-sized member of the sandpiper family, is one of the most northerly breeding species. It nests mainly on islands and

Below Winter-plumaged knots mass at high tide, waiting impatiently for the opportunity to resume feeding.

MIGRATION PROFILE

Scientific name	*Calidris canutus*
Migration	From Arctic breeding areas to temperate and tropical wintering areas
Journey length	2,500–16,000 km (1,500–10,000 miles) each way
Where to watch	Delaware Bay, New Jersey, USA; Morecambe Bay, England
When to go	Late May (Delaware Bay); November–February (Morecambe Bay)

Above With its long bill, legs, and toes, the knot is a typical wader. This colourful individual is in breeding dress.

peninsulas in the far north, between the isotherms of 1°C (34°F) and 5°C (41°F). As soon as the winter ice has melted, the female knot lays her clutch of three or four eggs in a scrape on the bare earth. The young are fully independent 39–42 days later, meaning that this species often leaves its breeding grounds within two months.

GLOBE-TROTTING JOURNEYS

Considering how far north the knot breeds, one might expect it to migrate to the nearest ice-free areas for the winter. In fact, it travels an extremely long way, sometimes to the other side of the globe. Many knots winter well south of the equator, flying as far as the coasts of Argentina, South Africa, Australia, and New Zealand; the rest spend the winter on seashores in the Caribbean, northwest Europe, and West Africa. By spreading out across the planet's temperate and tropical regions, the species is able to exploit a wider range of locations and thus reduce competition for food.

The knot, in common with other waders, minimizes the energy cost of its epic migrations in a variety of ways. First, it travels in compact flocks, which is more efficient; one radio-tracking study proved that flocks of knots could increase their average speed by 5 kph (3 mph) compared to a bird flying on its own. Second, the species takes advantage of tailwinds; it has been recorded flying at an altitude of 5,000 m (16,500 feet), where powerful winds could comfortably double its ground speed. Finally, it breaks its journey into stages, making two or three non-stop flights with pauses at productive refueling areas, such as sheltered estuaries. These "staging posts" are crucial to the knot and their location has, over thousands of years, come to dictate the migration routes taken by the species.

TIDAL ROOSTS

Although the knot is thinly scattered on its breeding grounds in the tundra, the reverse is true at its staging posts and winter quarters, where it forms huge concentrations to feast on small mud-living molluscs and crustaceans revealed by the tide. For example, 350,000 knots assemble each winter on the Banc d'Arguin, an area of sand banks off the coast of Mauritania, with up to 100,000 in the Wadden Sea along the Netherlands coast and another 300,000 divided between several English estuaries.

The birds feed at the water's edge, which means that the incoming tide forces them to bunch or "knot" together on raised banks, packed in so closely that they may land on top of each other. The massed ranks huddle for a couple of hours to sit out the high tide, but every now and again explode into the sky and wheel about in unison. When this happens, the air is suddenly thick with thousands of birds that twist and turn with an almost volcanic energy, as if one fluid organism.

Bar-tailed Godwit

Bar-tailed godwits make the longest non-stop flights of any land bird or wader. Members of the Alaskan population fly over the Pacific Ocean to Australia and New Zealand, burning their own innards to fuel the epic 200-hour crossing.

The bar-tailed godwit is an elegantly proportioned wader with a long, slightly upturned bill, and in its breeding dress is a beautiful shade of rich cinnamon. It resembles the three other species of godwits, all of which are powerful migrants. However, none comes close to matching the transcontinental journeys this species undertakes to and from its coastal winter quarters and its nesting areas in the far north.

MIGRATION PROFILE

Scientific name	*Limosa lapponica*
Migration	From Arctic breeding areas to temperate and tropical wintering areas
Journey length	2,000–14,500 km (1,250–9,000 miles) each way
Where to watch	Firth of Thames, North Island, New Zealand; Banc d'Arguin National Park, Mauritania
When to go	October–February

FOUR OF A KIND

Every year the bar-tailed godwit flies north to the desolate tracts of tundra, swamp, and peat-moss that encircle the top of the planet and which in summer form part of the so-called "land of the eternal sun". The godwit does not breed throughout this circumpolar zone but in four discrete regions: Norwegian Lapland and the White Sea area of northwest Russia; north-central Siberia, particularly on and around

Below Adapted for speed and endurance flying, godwits can stay airborne for more than a week at a stretch. Even so, their long flights over the sea leave little margin for error.

Above The bar-tailed godwit's tundra nesting grounds are a challenging environment, but the plentiful food makes it worth braving the harsh conditions.

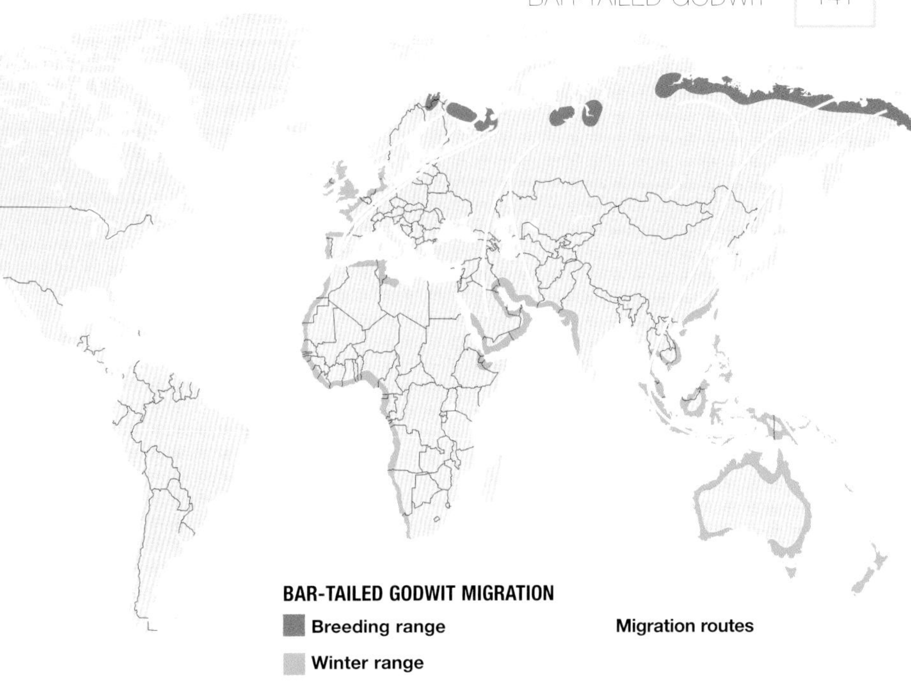

the Taimyr Peninsula; northeastern Siberia; and western Alaska. The birds originating from each area differ in size and have separate wintering grounds, so ornithologists regard them as biologically distinct subspecies, or races.

When it comes to migratory prowess, the four subspecies are far from equal. Most Scandinavian birds (*Limosa lapponica lapponica*) travel to the coasts of western Europe and South Africa, whereas the central Siberian birds (*L. l. taymyrensis*) fly to the sand flats of Banc d'Arguin in West Africa, the east Siberian birds (*L. l. menzbieri*) fly mainly to Southeast Asia and Australia, and the Alaskan birds (*L. l. baueri*) go to Australia and New Zealand. The last of these journeys is not only five times longer than the first, but also almost entirely over water, thereby denying migrants the chance to rest and feed on the way. Unsurprisingly, individuals that belong to the Alaskan subspecies are larger and heavier than those in the Scandinavian subspecies.

SURVIVING THE JOURNEY

Non-stop migration is enormously demanding for any bird, and the maximum range that it can attain depends on the ratio between its average flight speed and the amount of fuel, in the form of water and fat, that it can carry. The bar-tailed godwit is ideally placed in this respect, because it is able to fly fast, at consistent speeds of 55–70 kph (35–45 mph), while carrying a heavy fuel load. It improves its already excellent flight performance ratio by riding a tailwind at high altitude, which shortens its flight time by up to a half.

Preparation for a long-range, non-stop flight is essential, and, in common with a marathon runner training for a race, takes a considerable amount of time. After breeding, a bar-tailed godwit is not in good enough condition to set off direct from its nest-site. Instead, it moves to an area of invertebrate-rich tidal mudflats to moult and reach its optimum pre-migration weight. Much of the species' Alaskan population, for instance, funnels through the Yukon–Kuskokwim Delta in west Alaska, one of the world's largest river deltas, where the birds feed furiously to replace their flight feathers and lay down fat reserves. Prior to departure, each bird's heart and pectoral muscles increase in size, while its gizzard, intestine, liver, and kidneys – organs of little use during the flight – shrink to compensate. In other words, the godwit jettisons non-essential parts of its metabolic machinery, converting them to other tissue.

Around 70,000–100,000 Alaskan bar-tailed godwits make the trip to Australasia each year, including both adults and recently fledged young. The ocean crossing is a tremendous achievement for an experienced adult, let alone a young bird hatched only two months previously, and although some juveniles fly south in flocks with the adults, others navigate there without any adult assistance. Their annual round-trip involves a total flight time of up to 500 hours, if all the stages are added together, which means these birds may spend over 15 per cent of their life migrating.

TRACKING STUDY REVELATIONS

A new generation of lightweight satellite tags with long-life batteries is providing astonishing data about the migratory feats of bar-tailed godwits that has pushed back the known boundaries of what birds can achieve. On 17 March 2007, a tagged female godwit with the code name E7 left New Zealand and flew 10,219 km (6,350 miles) non-stop to the Yalu Jiang River in China, near the North Korean border, where she arrived 7 days 13 hours later. E7 paused for five weeks to fatten up, then set off again on 1 May, traveling 6,459 km (4,013 miles) without stopping to reach the Alaska Peninsula on 5 May. Her ultimate destination was even further into Alaska, where she finally arrived on 15 May. At the end of August that year, E7 set a new record by flying at least 11,570 km (7,189 miles) non-stop from Alaska back to New Zealand, taking 8 days 12 hours.

Arctic Tern

Arctic terns circumnavigate the globe on their phenomenal migrations between northern latitudes and the edge of the Antarctic pack ice. Some experience two polar summers in a year and see more hours of daylight than any other creatures on the planet.

As befits true globetrotters, Arctic terns are the only birds routinely to occur on each of the Earth's seven continents (cattle egrets nest on six continents and have been seen on subantarctic islands, but those sightings concern vagrants, observed far from their usual range). Arctic terns breed right around the entire Arctic Circle, from Alaska east to Canada and Greenland and from Iceland through northwest Europe to the Svalbard islands and the northern coasts of Siberia. In late July and August, they set out from their breeding colonies to migrate south, travelling over either the Atlantic or the Pacific Ocean to reach the icy shores of Antarctica.

THREE PATTERNS

The terns follow one of three main migratory tracks on their southbound journey. Those that nest in eastern Canada and

Below Blocks of sea ice make convenient perches for the terns during their stay in Antarctic waters.

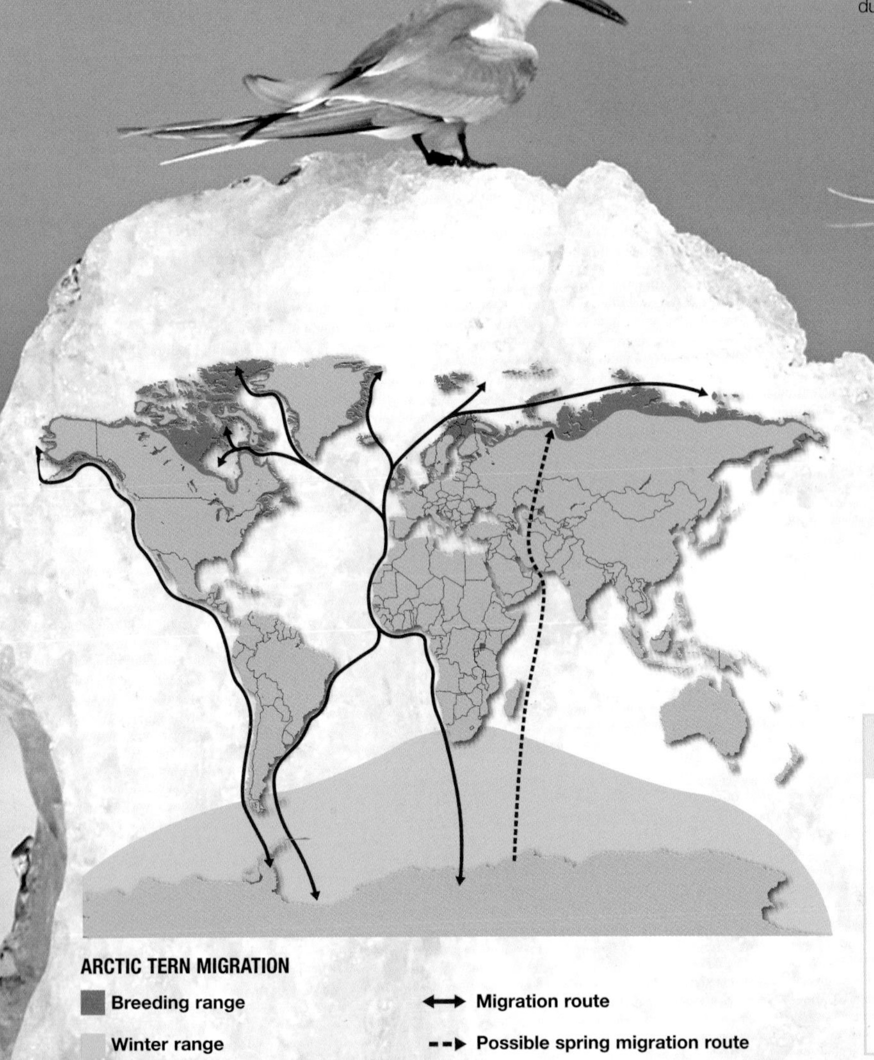

MIGRATION PROFILE

Scientific name	*Sterna paradisaea*
Migration	Global circumnavigation from pole to pole
Journey length	15,250–20,000 km (9,500–12,500 miles) each way
Where to watch	Coasts of Alaska, Canada, and northern Europe
When to go	May–July

LIFETIME MILEAGE

Arctic terns breeding in the far north are estimated to make a round-trip of about 40,000 km (25,000 miles) every year, and some might clock up as many as 50,000 km (30,000 miles) annually. The oldest known Arctic tern, an individual that nested in the northeast USA and bore the band number 35.325864, was 34 years old, so it follows that a tern of similar age nesting in the High Arctic could potentially have travelled up to a million miles.

Greenland begin by flying southeast across the Atlantic to join those moving south from Siberia and Europe. The passage then continues south along Africa's western coastline, with most birds advancing through African coastal waters as far as the Cape of Good Hope, before crossing the Southern Ocean to Antarctica. But some terns take a different route when they hit West Africa, by peeling off to traverse the Atlantic to Brazilian waters, from where they push south again, past Argentina and Patagonia. The terns of Alaska and northwest Canada form a third group. They skirt the entire Pacific coastline of North and South America, to Cape Horn at the southern tip of the Tierra del Fuego archipelago.

Once in Antarctic waters, the terns spread out and drift gradually south as they follow the retreating pack ice. The boundary where pack ice meets open water is highly productive, supporting colossal shoals of Antarctic krill (see pages 106–107) and small fish. Some terns are thought to complete a full circuit of the Antarctic continent during their few months spent harvesting this prolific resource, but finally, in early March, they head north again. Adults of breeding age return all the way to their colonies, while immatures pass the first two or three years of their life in the Southern Hemisphere.

SUN SEARCHERS

It has often been said that it would be difficult for Arctic terns to migrate further, even if they tried. Their pole-to-pole migration requires a huge energy expenditure, yet is worth it for the chance to benefit from two long summers a year. The areas in which they nest receive an average of between 18 and 24 daylight hours in summer, depending on latitude; in little over two months, the terns are able to pair up, establish and defend territories, raise young, and return to their seafaring life. Adults and juveniles travel south together.

Arctic terns are nicknamed "sea swallows" after their long tail streamers and dipping, swooping flight style, and as they float effortlessly above the waves are the epitome of grace, but there is mounting evidence that at least some of them regularly venture far inland. There are spring records of Arctic terns from Central Asia and the Ural Valley in Russia, which suggests that they use prevailing winds to migrate north through the Indian Ocean, and from there across Asia to their Arctic nesting areas. The terns probably fly very high, so are encountered only when they rest at suitable fresh waters en route. A variety of other seabirds, including skuas and murrelets, have also been recorded migrating through the centre of large land masses.

Above A breeding adult displays the aerial mastery for which the species is famous.

Below Arctic terns breed on shingle beaches or coastal grassland, often on offshore islands. Their nest is simply a shallow scrape in the ground.

Arctic Invasion

Birds from the far north sometimes stage mass invasions of lands to the south, driven by the failure of their usual food supply. These sporadic movements are carried out by birds of prey, owls, and a colourful array of seed and berry eaters.

The forests and bogs of the taiga and tundra zones are home to a diverse selection of birds that could be described as "reluctant" migrants. In normal years, they stay in fairly defined areas or make short journeys to escape severe winter weather. Occasionally, however, they are hit by chronic food shortages at the end of the breeding season when populations are highest, precipitating a crisis. Without warning, huge waves of birds evacuate northernmost latitudes, and move hundreds or thousands of miles in a generally southerly direction throughout the autumn.

Birds that "erupt" like this rely on a specific food source with a natural tendency to wax and wane. Irruptive species include, for example, rough-legged buzzards, snowy hawks, and great grey owls – all expert hunters of lemmings and voles. Numbers of the rodents plummet every three to five years, triggering a southbound invasion of hungry hawks and owls.

A second group of irruptive species comprises birds that depend on the seeds and berries of pine, spruce, birch, and mountain ash, which crop heavily for several seasons, then rest the following year. Birds in this category include members of the crow family, such as nutcrackers and jays; waxwings – starling-sized birds of boreal forests; and various finches.

Facing page When lemmings are scarce, snowly owls abandon their Arctic range and head to milder climes in search of alternative prey.

Above Evening grosbeaks have a powerful bill for cracking large seeds, and during invasion years they often take nuts and seeds from bird feeders in gardens.

Inset left In North America, northern goshawks move south when their favourite quarry, snowshoe hares, become harder to find. This happens about once every 10 years.

Inset left below Waxwings erupt southwards when berries run out in the north. They stop wherever they find new supplies, strip the trees bare, then press on.

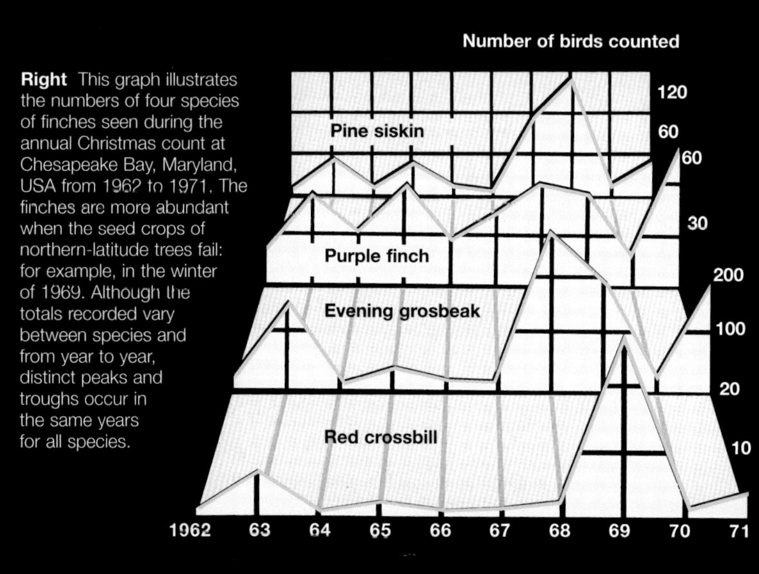

Right This graph illustrates the numbers of four species of finches seen during the annual Christmas count at Chesapeake Bay, Maryland, USA from 1962 to 1971. The finches are more abundant when the seed crops of northern-latitude trees fail: for example, in the winter of 1969. Although the totals recorded vary between species and from year to year, distinct peaks and troughs occur in the same years for all species.

Ruby-throated Hummingbird

The ruby-throated hummingbird's fast-paced, nectar-fuelled lifestyle depends on there being plenty of flowers, which forces it south in the autumn. When it returns in spring this tiny dynamo is capable of flying non-stop across the Gulf of Mexico – an incredible journey for a bird that weighs less than an everyday pencil.

Above Ruby-throats depend on about 30 species of flowers during the spring and summer. Red, pink, and orange blossoms seem to have a special attraction for them.

RUBY-THROATED HUMMINGBIRD MIGRATION

- Breeding range
- Winter range
- → Autumn migration route
- → Spring migration route

MIGRATION PROFILE

Scientific name	*Archilochus colubris*
Migration	From North America to Central American wintering areas
Journey length	Up to 6,000 km (3,750 miles) each way
Where to watch	Gardens and woodland in eastern North America
When to go	April–July

Above The female builds a neat, cup-shaped nest from soft plant fibres, lichen, and spider silk, and raises the two young entirely on her own. They fledge after 18–20 days.

Many of the world's approximately 340 species of hummingbirds live in the rainforests of Latin America, where they benefit from year-round nectar and rarely leave the same small patch of forest. Elsewhere in the Americas, hummingbirds face a problem when seasonal nectar supplies in their habitat dwindle and can no longer support their hyperactive lifestyle. A few species, such as the long-tailed sylph from the Andes, move downhill to lower levels where the warmer climate provides more food, but the rest migrate.

Of the 13 or 14 hummingbird species that regularly breed in Canada and the USA, all apart from a couple in southern California are strongly migratory. In the autumn they travel either south to Mexico and Central America, like the ruby-throated hummingbird, or east to the Gulf States and North and South Carolina.

FLOWER POWER

The ruby-throated hummingbird breeds throughout the eastern half of North America, north to a line from central Alberta to Nova Scotia, and is the only hummingbird that nests east of the Mississippi River. Males, resplendent with a glittering scarlet gorget and bronze-green upperparts, typically set off from their tropical winter quarters one or two weeks before the plainer, pale-throated females, giving them time to claim a nesting territory. They move through the southeast USA in late March and early April when three of their favourite food plants – cross vine, red buckeye, and firebush – are blossoming, and their progress north through the continent keeps pace with the flowering of other important food plants.

During the nesting season the ruby-throats' diet is often more varied, featuring tree sap sipped from bark, spiders plucked from webs, and small insects caught in mid-air or taken from flowers. Like all hummingbirds, the females raise their young without the help of their partners; nevertheless, a growing proportion of the population now produces two broods, probably due to longer summers and the popularity of garden bird feeders. Since they play little part in the breeding effort, male ruby-throats are ready to head south again as early as July, but the females and young remain in the breeding areas until September, or even mid-October if the weather is mild enough.

The ruby-throats follow a variety of routes south. At peak periods large numbers pour through major migration flyways, including the Mississippi Valley, Gulf coast, and eastern Texas, which lie at the crossroads between the temperate and tropical worlds. From Texas, they push on through Mexico and Central America, with some travelling all the way to Panama.

DANGEROUS CROSSING

Some ruby-throats retrace their overland routes as they return north in spring, but the rest make a shorter transit of the Caribbean – a journey of about 1,100 km (700 miles). This sea crossing would be too risky to attempt in the autumn, when hurricanes present a severe danger; even with a helping tailwind, it is still a phenomenal performance. In calm conditions it takes the birds about 18 hours and requires an estimated 3.2 million wingbeats at the rate of 49–50 per second. Relative to body size, it is one of the longest known non-stop migratory flights of any species of bird.

Before starting this supercharged marathon, the hummingbirds gather on the forested shores of the Yucatán Peninsula to put on the substantial fat reserves they will need as fuel. Their predeparture weight is roughly double their normal weight of 3 g (0.1 oz). Virtually all of the stored fat will have been burned to power the birds' huge pectoral muscles by the time they make landfall on the stretch of coast between Louisiana and Florida.

CHANGING PATTERNS

Hummingbirds, in common with humans, much prefer complex sugars, so these thirsty sprites adore the sucrose that we refine from sugar cane. Thousands of gardens scattered across North America boast sugar-water feeders, creating an extensive network of refuelling stations for migrating ruby-throats, as well as rufous hummingbirds and several other species. Hummingbird feeders are a particularly valuable resource in "food deserts" such as cities, which the migrants would otherwise have to detour around or fly over without stopping. Increasing numbers of hummingbirds are now adapting their migrations to winter in gardens in the southeastern USA, where the never-ending supply of synthetic nectar sustains them until spring. So far only a handful of ruby-throats appear to have followed this trend, which could in part be caused by global warming, but more may do so in future.

Above Garden sugar-water feeders are a lifesaver for migratory birds.

Southern Carmine Bee-eater

Resplendent in hot pink and adorned with elegant tail streamers, this is one of the most handsome birds of the African bush. The bee-eater is just as impressive on the wing, swooping effortlessly to intercept its insect prey in mid-air. It has a complex three-part migration, and travels far and wide through the savanna.

The southern carmine bee-eater, like most of the bee-eater family, is a highly social species that nests in earth banks in tightly packed rows of burrows. Its colonies usually have between 100 and 1,000 occupied nest-holes as well as many disused burrows, but in a good year some colonies contain up to 10,000 active nests. An ideal breeding site, such as a high sandy bank overlooking a meandering river or oxbow lake, will be used by the bee-eaters for many years, becoming so heavily pockmarked that it gives the impression of having been peppered with machine-gun fire. Eventually the birds' tunnelling weakens the cliff and it crumbles away, forcing the colony to relocate a few miles up or down river.

With the endless coming and going of thousands of brilliantly coloured birds, a large carmine bee-eater colony ranks among the top birdwatching spectacles in Africa. But when the bee-eaters finish breeding and move on, the nesting bank may be deserted for eight months.

TROPICAL MIGRANT

The southern carmine bee-eater is almost entirely an intra-tropical migrant – that is, its seasonal movements are within the tropics. It breeds in a broad belt of land stretching almost across south-central Africa, from Angola, east through Zambia and the Okavango Delta in northern Botswana, to Zimbabwe and Mozambique. This range corresponds to a vegetation zone dominated by savanna and dry deciduous forest. At the heart of this huge area, near the southern end of the Great Rift Valley, is Zambia's Luangwa River – one of Africa's greatest seasonal rivers and a breeding stronghold for carmine bee-eaters. Here, the first migrant bee-eaters arrive in July each year, but the main influx of birds takes place in August. It is suspected that the bee-eaters return to the same breeding cliff as the previous year, although some colonies are more mobile than others, shifting location most seasons.

FIRE BIRDS

In West Africa, the Mandinka people call the northern carmine bee-eater "cousin to the fire". This wonderful name describes a fascinating aspect of its behaviour, and suits its southern relative equally well. Both species often fly close to grass fires, at times dipping to within a few feet of the blaze. They are drawn to fires by the fleeing hordes of locusts, grasshoppers, and beetles driven into the air by the rapidly moving wall of flames, making easy pickings.

Each pair of bee-eaters digs a new burrow, rather than cleaning out an old one, with the result that crowded colonies reach a high density of about 60 nest-holes per square metre. There is fierce competition for the best spots on the cliff – neighbouring pairs engage in long aerial chases and a mid-air tussle occasionally brings rivals tumbling to the ground.

MULTI-STAGE MIGRATION

In Zambia most carmine bee-eaters lay eggs in September, towards the end of the April–November dry season, so that their young are ready to leave the nest during the onset of the wet season when insect food is plentiful. By December the adult bee-eaters and juveniles have already begun to disperse. Most of them appear to spread south into higher latitudes, moving as far as the Transvaal in northeast South Africa – a distance of at least 650 km (400 miles). The bee-eaters stay here for up to three months, until March, and then embark on the next stage of their migration. Slowly they head northwards again, this time travelling over their breeding range to reach the savanna at low latitudes, sometimes nearing the equator. They are joined by the birds that did not originally fly south and instead lingered in the breeding area.

Outside the breeding season, carmine bee-eaters are mainly nomadic, ranging widely in loose flocks. They take advantage of locally abundant food sources – a swarm of honeybees, insects disturbed by a fire or herd of grazing animals, a synchronized hatching of bugs or termites, or a mass movement of desert locusts (see pages 164–165).

Below Flocks of bee-eaters are strongly attracted by the sight of smoke, and perhaps also by the distant roar of the inferno itself.

Above The nesting cliffs used by carmine bee-eaters are transformed into a riot of colour as birds stream in with beakfuls of food and bicker with their neighbours.

Barn Swallow

Long associated with the arrival of spring and celebrated as heralds of good weather, swallows are among the Northern Hemisphere's best-loved migratory birds. Completely at home in the air, they migrate at a leisurely pace, feeding as they go.

Barn swallows, known simply as "swallows" in the UK, are delightful birds and truly cosmopolitan, the most widespread species of swallow in the world. They breed in open country throughout the Northern Hemisphere, except for Arctic regions, and winter across the Southern Hemisphere, apart from the arid lands of Australia and northern Africa. Their attachment to lush, cattle-grazed pasture and habit of nesting on artificial structures bind their fortunes to human modification of the landscape; in return, their vast appetite for insect pests makes them a great friend of agriculture.

The symbolic link between barn swallows and spring in Western culture dates back 2,500 years to Ancient Greece, and in the fourth century BCE. Aristotle became the first writer correctly to identify them as migratory. In medieval times, swallows were widely believed to spend winter hibernating in mud at the bottom of ponds, a fantastic theory that persisted into the late 1700s, but we now know from ringing studies that some of the old adages surrounding this species are based in fact. Barn swallows are indeed reliable portents of spring, and they are faithful too, with a strong tendency to return to the same nesting area. In one North American study, 65 per cent of adult swallows came back to claim the previous year's nest or another next to it. In another experiment, a swallow captured at its nest and released 1,725 km (1,070 miles) away managed to navigate "home" with apparent ease.

FAIR-WEATHER EMBLEM

Barn swallows depend on an invisible habitat – the shallow layer of sun-warmed air above ground level – and the insect hordes that

Below Barn swallows are wonderfully elegant, aerobatic birds that even drink on the wing, dipping to the surface of ponds and lakes to scoop up billfuls of water.

sustain them vanish almost overnight in cold conditions. As a result, their movements match changes in the prevailing air temperature. Swallows returning north from their winter quarters travel more slowly and arrive later in cold springs than in warm ones, for instance. Likewise, in a warm autumn the birds delay their southbound migration to take advantage of the unseasonal weather.

The swallows breed across such a wide span of latitude that each spring it takes them over three months to recolonize every part of their breeding range. In the south, swallows start arriving in March, giving pairs ample time to raise three broods of young, while the northernmost contingent, which nest in Alaska, Scandinavia, and Siberia, do not arrive back until late May or June and produce only a single family. Not all barn swallows are migratory, however. There are sedentary populations in central Mexico, southern Spain, and Egypt, which stay put as hundreds of thousands of swallows from distant migratory populations overfly their areas.

DAY BY DAY

Unlike most small, insectivorous songbirds, barn swallows are mainly daytime migrants and thus can feed on the wing as they travel (the chief exception to this rule occurs during their non-stop transits of large deserts, such as the Sahara, which are made in the cool of night). Each day the swallows set off shortly after dawn, and in the evening they converge on large roosts hundreds-strong, most often in dense reedbeds over water for protection from ground-based predators. Some birds, especially immatures, will use a particular roost for up to a couple of weeks before pushing on, which means that the swallows' north–south passage is rather slow compared to those of other long-distance migrants.

Ringing data indicate that swallows from northern Europe take approximately 10 weeks to complete their journey to their wintering areas in southern Africa. They cover a daily average of 150 km (95 miles), but this figure is misleading, since the swallows migrate in fits and starts, taking a break between each flight. There is a greater urgency about the northbound passage in spring, which, driven by the need to reproduce, is roughly twice as fast, taking five to six weeks.

Barn swallows frequent wetlands and savanna in their winter range, where their overnight roosts are often massive. In the late 1990s, a cluster of 34 acacia trees in Botswana regularly held a roost of a million swallows (almost 30,000 birds per tree), while an astonishing five million swallows – over eight per cent of the European breeding population – use Mount Moreland reedbed near Durban, South Africa. The latter roost site was threatened by a new airport development but won a reprieve in 2007.

Below This flock of swallows has paused to rest in an acacia tree while on migration through the savanna of East Africa.

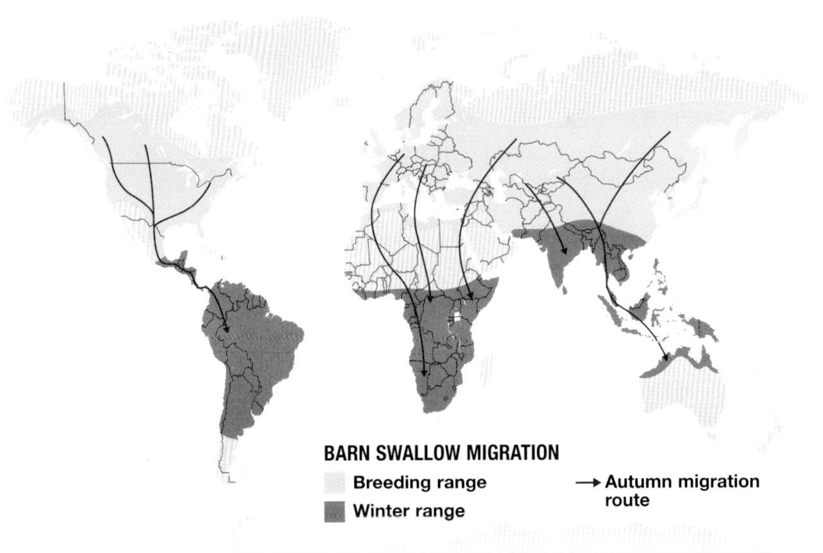

RESTLESS GATHERINGS

Twittering flocks of barn swallows lined up on telephone wires or power lines are a sure sign of the end of summer. There is an air of expectation about these annual gatherings: the birds flutter their wings rapidly; switch position; and chase each other or circle overhead in small, excited groups. This behaviour is one of the best-known examples of zugunruhe, the restlessness of migratory birds prior to departure. Diurnal migrants such as barn swallows display zugunruhe by day; nocturnal migrants such as thrushes, wood-warblers, and Old World warblers become restive at night.

MIGRATION PROFILE

Scientific name	*Hirundo rustica*
Migration	From Northern Hemisphere breeding grounds to wintering areas in Southern Hemisphere
Journey length	Up to 12,000 km (7,500 miles) each way
Where to watch	Farmland in USA, Canada, and Europe
When to go	April–August

Willow Warbler

One in five of all the migratory birds that pour out of Europe and Asia to winter in Africa's warmer climes are willow warblers – a post-breeding population of nearly a billion adults and juveniles. This flood of tiny birds surges south for thousands of miles, crossing mountains, seas, and desert.

The willow warbler belongs to the Old World warblers, a large family famous for its long-haul migrants, and is not related to the wood-warblers of the Americas (see pages 156–157). It is named for its fondness of willow glades and is the commonest bird throughout much of its breeding range, which stretches between the July isotherms of 10°C (50°F) and 22°C (72°F), north into the Arctic tundra. In its African wintering grounds, the willow warbler frequents virtually any kind of country with trees, including acacia savanna and evergreen forest.

A traditional harbinger of spring, the willow warbler's soft, quavering song floats across vast tracts of Europe and Asia in April and May as males establish their nesting territories and compete to attract a mate. When one male begins to sing, it triggers an immediate answer from its neighbours and the wood or heath comes alive with a chorus of song. The species is a rapid breeder, on average raising a brood of four to eight young in 26–28 days, so is ready to embark on its southbound migration from late July to August.

DIFFERENT JOURNEYS

The willow warbler's migrations are well known, due to the sheer number of individuals ringed every year. By 2004, British ornithologists had caught more than a million birds, of which 2,500 were subsequently retrapped or picked up dead. This is a good recovery rate in ringing terms, providing a substantial data pool.

Above Within a few months, the survivors from this young brood of willow warblers will have set off to fly to Africa.

Ring-recoveries show that birds from the west of the species' European range travel on a south to southwest heading in the autumn, via France and Spain, to spend the winter in West Africa. Birds from north and east Scandinavia set off in the opposite direction: they fly south to southeast, wintering in Central, East and southern Africa. The population that breeds in Siberia has to fly the furthest. These birds head first south and then increasingly southwest across Russia, mostly via the Ural Valley, and probably travel all the way to southern Africa, a marathon journey of at least 14,000 km (8,700 miles). Depending on their origins, the warblers arrive in Africa between September and December. While in their wintering areas they lead an itinerant existence, wandering from place to place in loose flocks with local birds.

FLIGHT STRATEGIES

Measurements taken from trapped willow warblers have given an insight into how such small birds, weighing as little as 8–12 g (0.25–0.4 oz), manage to complete their arduous migration. These night migrants appear to have a fast refuelling capacity and therefore can use their daytime stops to feed en route. The hot sands of the Sahara Desert present the migrants with a significant barrier, but before the warblers cross they quickly put on enough fat to make the journey to the other side: migrating warblers trapped in Egypt in the autumn had sufficient body fat to cross the desert in three nights, resting during the two intervening days.

As with many migratory birds, the timing of the migration in willow warblers varies according to sex and age. In the fall juveniles migrate south before the adults, which catch up or overtake them later. The earlier departure of young birds could be explained by their slightly different structure – their rounder wing profile compared to the longer-winged adults means that they are less efficient at flying long distances. Meanwhile, on the return spring migration, males move through Europe about two weeks before females, suggesting that they start their migration in Africa sooner. This head start gives them long enough to form territories by the time their potential partners arrive, and in places they may become the most abundant species.

There is growing evidence that climate change is leading to shifts in this established pattern. For example, British willow warblers, in response to warmer spring weather, lay their eggs a week earlier than they used to, and leave the UK later than they did 40 years ago.

Right Freshly arrived on his nesting territory, a male willow warbler delivers his languorous, melancholic song. He will sing for several hours a day.

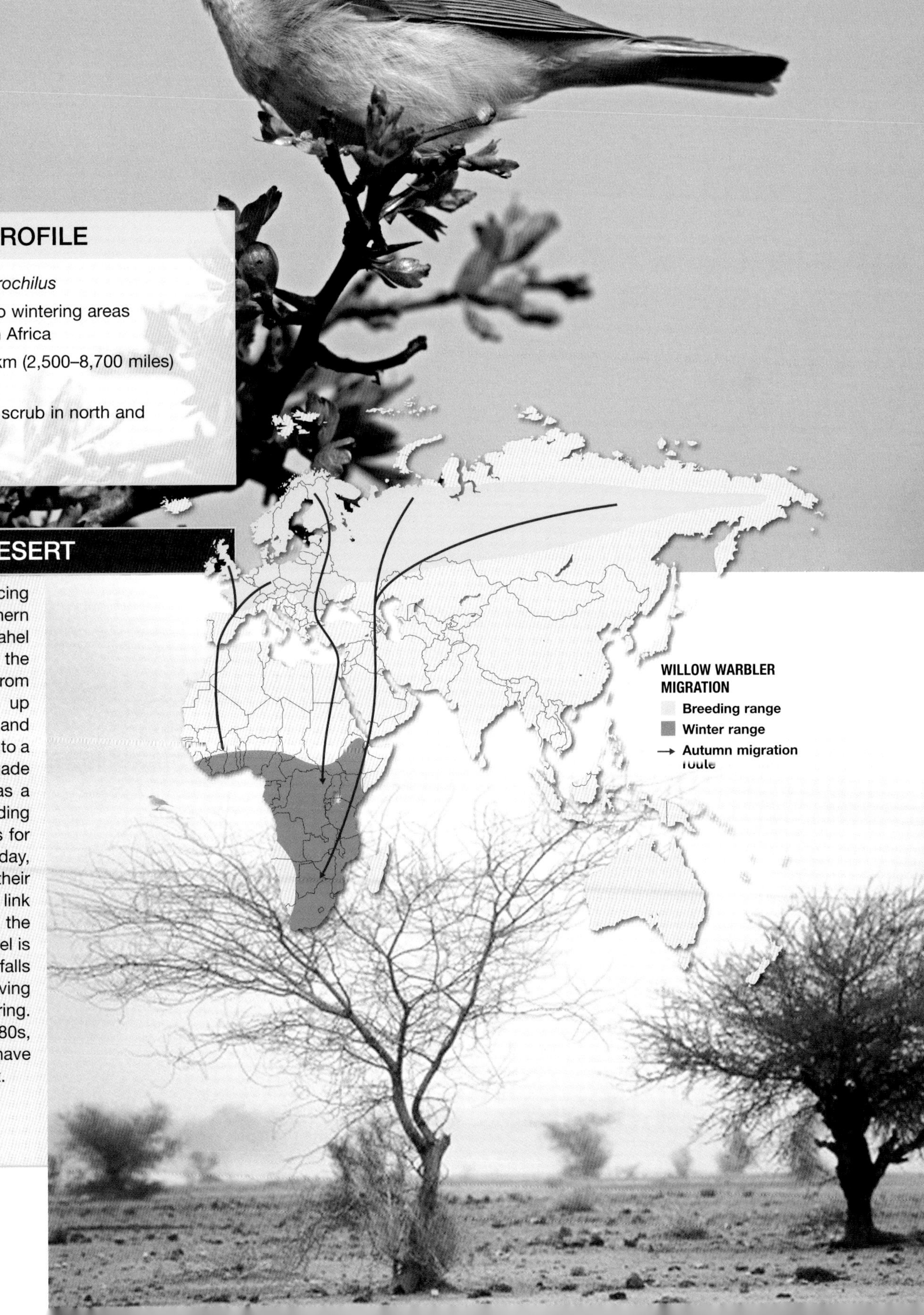

MIGRATION PROFILE

Scientific name	*Phylloscopus trochilus*
Migration	From Eurasia to wintering areas in sub-Saharan Africa
Journey length	4,000–14,000 km (2,500–8,700 miles) each way
Where to watch	Woodland and scrub in north and central Eurasia
When to go	April–May

SPREADING DESERT

Possibly the most severe threat facing willow warblers is drought in the northern half of Africa. Desertification in the Sahel region, which forms a band along the southern edge of the Sahara Desert, from Senegal east to Sudan, has dried up once-reliable sources of fresh water and turned formerly green areas of bush into a barren, dusty plain. The situation is made worse by chronic overgrazing, and as a result, the Sahara is effectively spreading south. This reduces the opportunities for tired migrant birds to rest during the day, while increasing the overall length of their trans-Saharan journey. Although the link has yet to be proved conclusively, the environmental catastrophe in the Sahel is likely to be a major cause of recent falls in the number of willow warblers arriving back on their breeding grounds in spring. A rapid decline set in during the 1980s, and some European populations have since fallen by more than 30 per cent.

Right The relentless expansion of the Sahara Desert has been blamed for a decline in the willow warbler population.

MIGRATION PROFILE	
Scientific name	*Ficedula hypoleuca*
Migration	From Eurasia to West African wintering areas
Journey length	2,800–7,250 km (1,750–4,500 miles) each way
Where to watch	Open woodland in Europe
When to go	April–May

Pied Flycatcher

Over thousands of years the northbound migration of pied flycatchers evolved to coincide precisely with the emergence of caterpillars in European forests. Now global warming has upset this age-old relationship, and the flycatchers' migration is increasingly out of sync with the caterpillar crop.

Pied flycatchers belong to the family Muscicapidae, or Old World chats and flycatchers, which comprises about 275 species found in Europe, Asia, and Africa. They are named for the striking black-and-white breeding plumage of adult males (females and non-breeding males are plain brown and white), and are active, restless birds, with a distinctive habit of flicking their tail and wings when perched. They seldom stay still, frequently swooping to pluck a bug from a leaf or darting out into the open to snatch a passing fly. In spring and early summer caterpillars are their chief food – the nestlings are fed on little else. Reproductive success in these handsome woodland songbirds is therefore closely linked to caterpillar abundance.

Facing page Male flycatchers arrive on the breeding grounds first, to claim territories before the wave of females arrives. Multiple partnering and cuckoldry is commonplace.
Above Most of the females end up as single parents. Somehow each manages to raise a brood of up to five nestlings with no help from the absent father.

HOME SWEET HOME

The male pied flycatcher is often polygynous, dividing his time between two or even three females, each of which nests in a separate territory. The male may therefore have to defend several territories against intruders, including other hole-nesting birds as well as rival males. He may abandon the first mate to spend all of his time with the second, or, after initiating the second brood, return to his original mate to help her. In any case, the females undertake most of the work, even when the male flycatcher stays faithful to one partner.

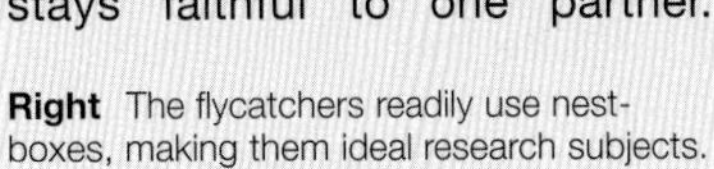

Right The flycatchers readily use nest-boxes, making them ideal research subjects.

CATERPILLAR FEAST

The annual explosion in caterpillar numbers varies with latitude across the flycatchers' large breeding range, which incorporates wooded areas from Spain east through central and northern Europe to southern Siberia. Clearly, it is advantageous for nesting flycatchers to choose an egg-laying date so that, when their chicks hatch 13–15 days later, this bonanza of food is waiting to be exploited. To achieve this match, the birds must time their northward migration from their winter quarters in tropical West Africa with great accuracy.

When in Africa the flycatchers cannot predict the advancement of spring in their breeding habitat – how then do they arrive on schedule? Evolution appears to have perfected their migration by selecting individuals that set off at the right time and travel at the right speed, though other factors may have played a part, too. Traditionally, pied flycatchers return north in March or early April, arriving in western Europe from mid-April to early May and reaching Scandinavia and Siberia in mid-May. As soon as the males are safely back on their breeding grounds, they begin singing to advertise territorial ownership.

OUT OF STEP

Since the 1980s there has been a trend towards warmer spring weather in Europe, which has resulted in caterpillars emerging earlier. In turn, pied flycatchers have started to lay their eggs earlier too. However, there is a limit to how soon the females are able to lay as they need to recover from their migration first. Moreover, the flycatchers appear unable to bring forwards their departure date from Africa by more than a week. One theory is that this is because their cue to leave depends on something unaffected by climate change, such as day length.

There is, however, firm evidence that global warming is causing the migration of pied flycatchers to be mistimed, with potentially devastating effects. In 2006, a study in the Netherlands reported a 90 per cent fall in flycatcher populations over the previous 20 years in areas where mistiming was worst. The species remains common overall, but localized declines in population could spread if the warming trend accelerates.

IBERIAN DETOUR

Pied flycatchers nest in holes, but readily use artificial nest boxes. This means that they are ideal research subjects, as ornithologists can observe the comings and goings at each numbered box. They are also a popular species for ringing studies, with over 450,000 individuals trapped and logged by 2004, and the high level of ring recoveries – nearly 3,500 to date – has revealed a wealth of information about the birds' migration strategies. One of the most intriguing aspects of their autumn migration is the use of a staging post, or refuelling area, in Portugal and northwest Spain. Most of the European breeding population passes through the cork oak forests of this region, which for eastern breeders involves a long detour west through the middle of the continent.

In much the same way that drivers interrupt a long journey to fill up with petrol, the flycatchers pause here to fatten up before the next stage of their trip. They defend separate feeding territories and gorge on berries as well as insects, increasing their body weight by up to a quarter. Having boosted their fat reserves, the flycatchers fly due south to the Strait of Gibraltar, where the sea crossing to Africa is narrowest.

Above Blackpoll warblers are well adapted for migration, with longer wings than sedentary birds of the same size. Their maximum lifetime journey is equivalent to an adult human travelling to the Moon and back ten times.

Below The autumn migration of blackpoll warblers across the West Atlantic coincides with the peak hurricane season, so there is always a risk that they will fly into the path of a tropical storm sweeping up the East Coast of the USA from the Caribbean.

TRANSATLANTIC VAGRANTS

If migrating warblers hit a hurricane, the spiralling winds, towering clouds, and torrential rain can spell disaster. Many birds crash into the sea exhausted, while those that stay airborne become disoriented. Sometimes the lost migrants are sucked into fast-moving depressions hurtling across the Atlantic and make landfall on the coasts of northwest Europe. They often turn up at the same time as other storm-blown American land birds, including robins, thrushes, and vireos. These vagrants rarely survive long and have no chance of making it back to the Americas, as prevailing winds in the Atlantic are easterlies and the birds' in-built migratory programme cannot guide them west across the ocean.

Blackpoll Warbler

During the fall, some blackpoll warblers make a marathon flight over the West Atlantic, travelling non-stop for more than three days. This dangerous ocean crossing is quicker than the overland route, but it requires phenomenal reserves of energy for these very small birds.

When the spring thaw sets in, the vast coniferous forests of the North American high boreal zone welcome a host of migratory birds from further south. This influx of summer visitors includes countless blackpoll warblers, one of the commonest breeding birds of the region's cool, damp spruce and fir forests. Like most of the other migrants, the warblers have flown north to take advantage of long daylight hours and the seasonal abundance of insect and spider prey, which enable them to raise a family rapidly.

In May and early June the newly arrived male blackpolls, wearing a distinctive black-and-white breeding plumage, perform a high-pitched, insect-like song to advertise their territories to the duller, greyish females. The species' breeding season is short: the female of each mated pair builds a cup-shaped nest of twigs and lichen in a small tree, and incubates her clutch of three to five eggs for 12 days, then both parents feed the young for another 12 days or so.

After breeding, blackpoll warblers remain in the northern forests to fatten up and moult their feathers, with the adult males adopting a relatively nondescript "autumn" plumage that closely resembles that of the females. Freshly moulted warblers begin to vacate their summer range during August, and the last few birds have left by the end of September, although there are signs that the migration is starting later in response to warmer weather in the autumn – this may become a long-term shift in behaviour, induced by climate change.

EAST THEN SOUTH

Blackpoll warblers are members of the American wood-warbler family, which is, confusingly, unrelated to the Old World warbler family (see pages 152–153). Like that group, however, the wood-warblers include many long-haul migrants, and the transoceanic journey undertaken by some blackpolls is arguably the greatest feat of endurance and navigation of any of the approximately 55 species of wood-warbler that regularly breed in the USA and Canada.

Blackpoll warblers winter in the tropical forests and shade-grown coffee plantations of northern South America, occasionally reaching southern Brazil, and the route they take to get there has been the subject of much debate and study. Instead of heading due south through the Midwest towards the Gulf of Mexico, as one might expect, most of the breeding population moves steadily southeast through the Great Lakes area to reach the coast of New England and Canada's Maritime provinces, where the birds linger a while. From here, many blackpolls follow the East Coast south, with some flying to Florida and then crossing the Caribbean and the others travelling only as far as North Carolina before heading out to sea to "island-hop" through the Bahamas and West Indies to South America. But the remainder avoid the East Coast altogether by flying directly out into the Atlantic. These birds travel southeast over the ocean until they pick up strong winds that drive them in a huge loop southwest towards the eastern Caribbean and Venezuela's north coast – a non-stop flight of 82–88 hours.

MIGRATION PROFILE

Scientific name	*Dendroica striata*
Migration	From North America to South American wintering areas
Journey length	4,000–8,000 km (2,500–5,000 miles) each way
Where to watch	Coniferous forest in Alaska and Canada
When to go	May–June

BLACKPOLL WARBLER MIGRATION
Breeding range
Winter range
Spring migration route
Autumn migration route

FLYING HIGH

Why do some blackpoll warblers carry out such a perilous migration over the Atlantic? The answer is that, due to the Earth's curvature, it is roughly 2,400 km (1,500 miles) shorter than the safer overland route. Routes that track the circumference of the planet in this way are known as "Great Circle" routes and are used by commercial airliners as well as birds. Aircraft fly high to cruise in the thinner air, and birds do this too. Blackpolls, for example, fly at altitudes of up to 5,000 m (16,500 feet), probably to reach the best tailwinds or maybe because the cooler air stops their hard-working breast muscles from overheating. In order to fuel this epic journey, the warblers feed intensively before setting off, and are able to boost their body weight to 20 g (0.7 oz) from an average 11 g (0.4 oz). All of this extra fat will be burned off during the flight – an achievement far beyond the capabilities of the human body.

In spring, blackpoll warblers follow different routes north. Most of them migrate northwards across the western Caribbean and then through the Mississippi Valley and Great Plains, taking about five weeks to fly from the Gulf Coast to Alaska.

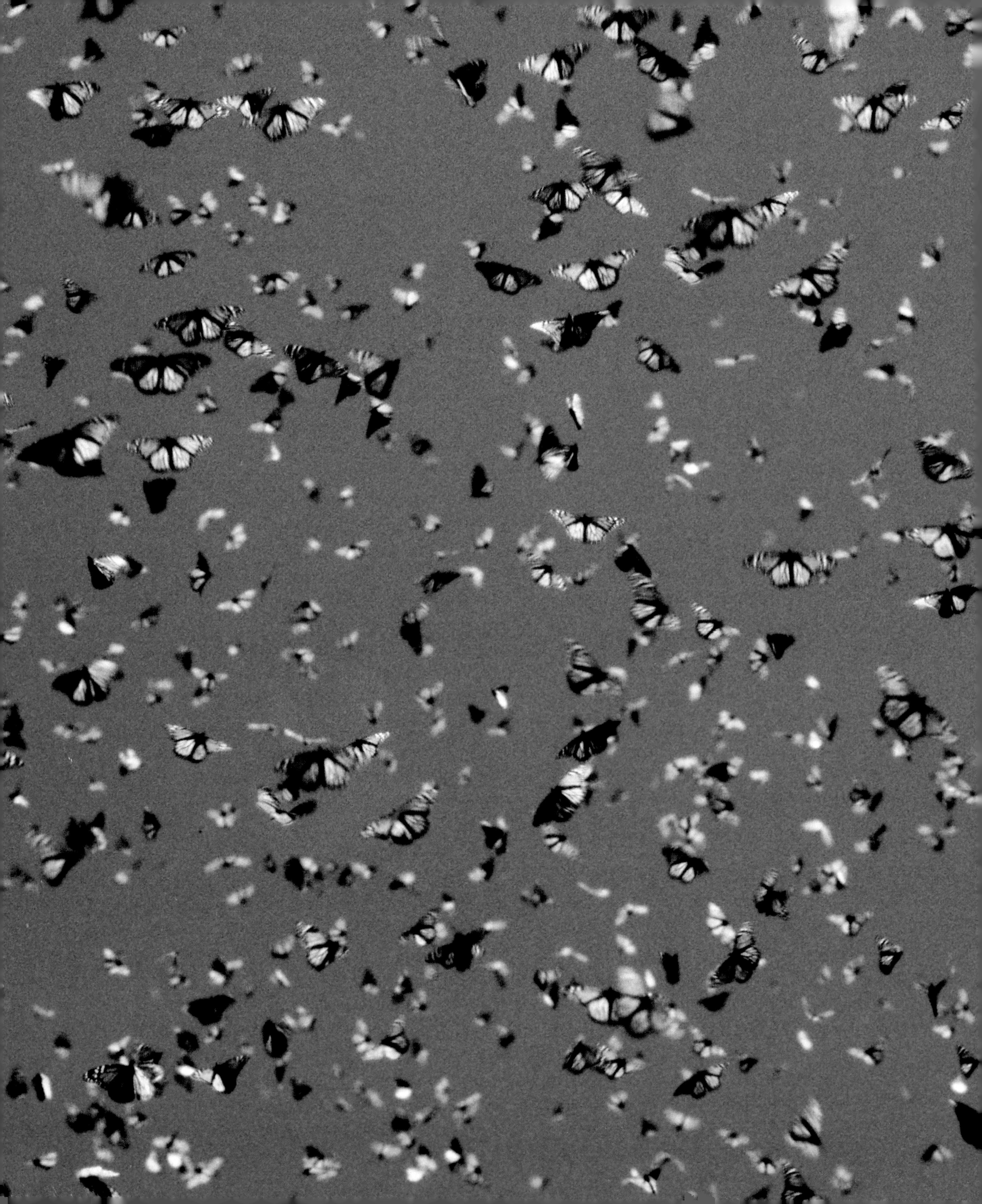

Hordes of Tiny Wings

Many insects are capable of long and complex migrations, yet their brains may be no bigger than the full stop at the end of this sentence. The secret of their success is to go with the wind and to travel in such prodigious numbers that some always make it.

A myriad of insects from temperate regions exhibit migratory behaviour, much of which remains mysterious or has come to light only recently. The champion migrants are naturally the strongest fliers – butterflies, moths, dragonflies, and grasshoppers – but plenty of other species undertake impressive journeys, including ladybirds, wasps, planthoppers, and even tiny aphids. Unlike vertebrates, which respond to a broad spectrum of migratory cues, insects are most sensitive to air temperature.

Because insects react to quite small changes in temperature, they are good indicators of the impact of global warming. In Britain, the peak summer arrival of painted lady butterflies and hummingbird hawkmoths, which travel north from wintering sites in the Mediterranean and North Africa, has crept forwards throughout the past two to three decades.

Insects are able to move extraordinary distances, crossing entire oceans on occasion. Migrating swarms of painted lady butterflies, for instance, regularly cover 2,000 km (1,250 miles) in under a month. Most insects fly slower than 3 m/sec (10 feet per second), so long-haul travel would be impossible without an external energy source, in the form of a prevailing wind or upwelling thermals. But riding weather systems is dangerous due to the risk of being blown off course, and many perish en route.

Above Migrant painted lady butterflies sweep north across Europe in summer. In some years, they reach the Arctic circle.

Left Butterfly wings are made of a carbohydrate called chitlin, which is strong, rigid, and ultra-light – the perfect combination for migratory flight.

Below, left to right Australian bogong moths migrate in large swarms, which gather on buildings to rest during the day; in higher latitudes, the key factor for insect migration is often the first hard frost in autumn and the last in spring; hummingbird hawkmoths are strong migrants that fly northwards after wintering in the Mediterranean and North Africa.

Monarch Butterfly

As summer draws to a close, more than 100 million monarch butterflies sweep across North America to overwinter in the pine forests of California and Mexico, returning north again in spring. These orange-tinted processions are one of the wonders of the insect world.

Monarch butterflies, like most insects of temperate regions worldwide, have had to evolve a system to cope with inhospitable wintry weather. Neither the adult butterflies nor their eggs, caterpillars, or pupal cocoons could withstand the subzero temperatures of a regular winter in Canada or northern and central parts of the USA. The monarchs' solution is an annual journey to safe roost sites further south, for which the trigger is probably the increasing length and coldness of nights in early autumn. Many other butterflies and moths, as well as some dragonflies and bugs, are medium- or long-range migrants, but the monarchs' migration is unusual because the southbound adults survive all winter to make the return trip.

FIVE GENERATIONS

The migration of monarch butterflies is best described as an intergenerational relay. In every breeding season up to five generations of butterflies live, reproduce, and die in North America, with the period between egg-laying and the emergence of a fresh adult from its chrysalis being 34–39 days, or less in warm weather. Adults on the wing in late summer behave differently from the others before them. They do not develop sex organs, and binge on nectar to lay down stores of fat, which eventually account for a third of their body weight.

At the end of August or in the first half of September, these monarchs – by now obese in butterfly terms – start to head south. They advance along a broad front in ever-increasing numbers, until finally the orange tide is tens of millions strong. We now know that the butterflies follow long-established routes, often tracking "leading lines" such as river valleys, coasts, or mountain ridges, and cover about 130 km (80 miles) on a typical day. However, the migration is a "stop–start" affair, with regular stops at nectar-rich flowers to top up fat reserves.

Each night swarms of monarchs rest communally in trees, using the species' traditional roosts, perhaps located by a faint trace of scent left by previous generations of migrants. The ultimate destination for around five million monarchs – those from the northwest USA and southwest Canada – is coastal California, where they settle in stands of eucalyptus trees and Monterey pines to sleep through the winter. The remaining population, which comprises at least 100 million individuals originating from the eastern half of Canada and the USA, converges on a dozen small sites in the pine and fir forests of central Mexico's volcanic mountain belt.

WINTER SLUMBER

Overwintering monarchs gather at such high densities that the branches of their trees bend with the sheer weight of encrusting butterflies. But this is no easy feast for birds and other insectivores: the dormant butterflies are left alone because their bodies are laced with toxins accumulated from the poisonous sap of their larval foodplant – milkweed. The butterflies do not wake until sometime in February or early March, and as clouds of them stir into life their wings whir audibly.

Within days, the monarchs mate, then head north. Those that wintered in California move to that state's Central Valley and the foothills of the Sierra Nevada, where they lay eggs and die, while Mexican populations fly to southern Texas to lay theirs. The adults in this generation are up to nine months old by the time they perish, making them some of the longest-lived of all butterflies. Later generations of monarchs move northwards and further inland throughout the spring and summer, leapfrogging one another as they reoccupy the species' full breeding range.

DISTANT COLONIES

Entomologists once calculated that if monarch butterflies flap their wings 8–10 times per second, they can travel 1,000 km (625 miles) non-stop under their own power in calm air, although the usual scenario is for them to climb high in the sky to take advantage of tailwinds that provide a free push. They also take advantage of boundary currents along rivers and coastlines. Occasionally in the autumn butterflies may be sucked into fast-moving weather systems that whisk them off-course to far-flung shores. These chance events are not necessarily disastrous for the species, as it has been able to colonize both Hawaii and Australasia in this way.

MIGRATION PROFILE

Scientific name	*Danaus plexippus*
Migration	To and from winter roost sites in south
Journey length	Up to 4,750 km (3,000 miles)
Where to watch	El Rosario Reserve, Michoacán, Mexico
When to go	February–early March

Above In California, where this photograph was taken, and also in Mexico, the celebrated overwintering monarch butterflies have become a major tourist attraction. They draw daytrippers from far and wide, as well as growing numbers of foreign tour groups. Recognizing the value of ecotourism, the Mexican authorities have created several butterfly sanctuaries, such as the one at El Rosario.

Inset Adult monarch butterflies feed intensively during late summer and early autumn, laying down energy reserves that will provide fuel throughout their long migration and last them through the coming winter. Swarms of the hungry insects seek out nectar-rich flowers, smothering the blooms as they drink their fill. By the time they finally set off, their tiny bodies are packed with stored fat.

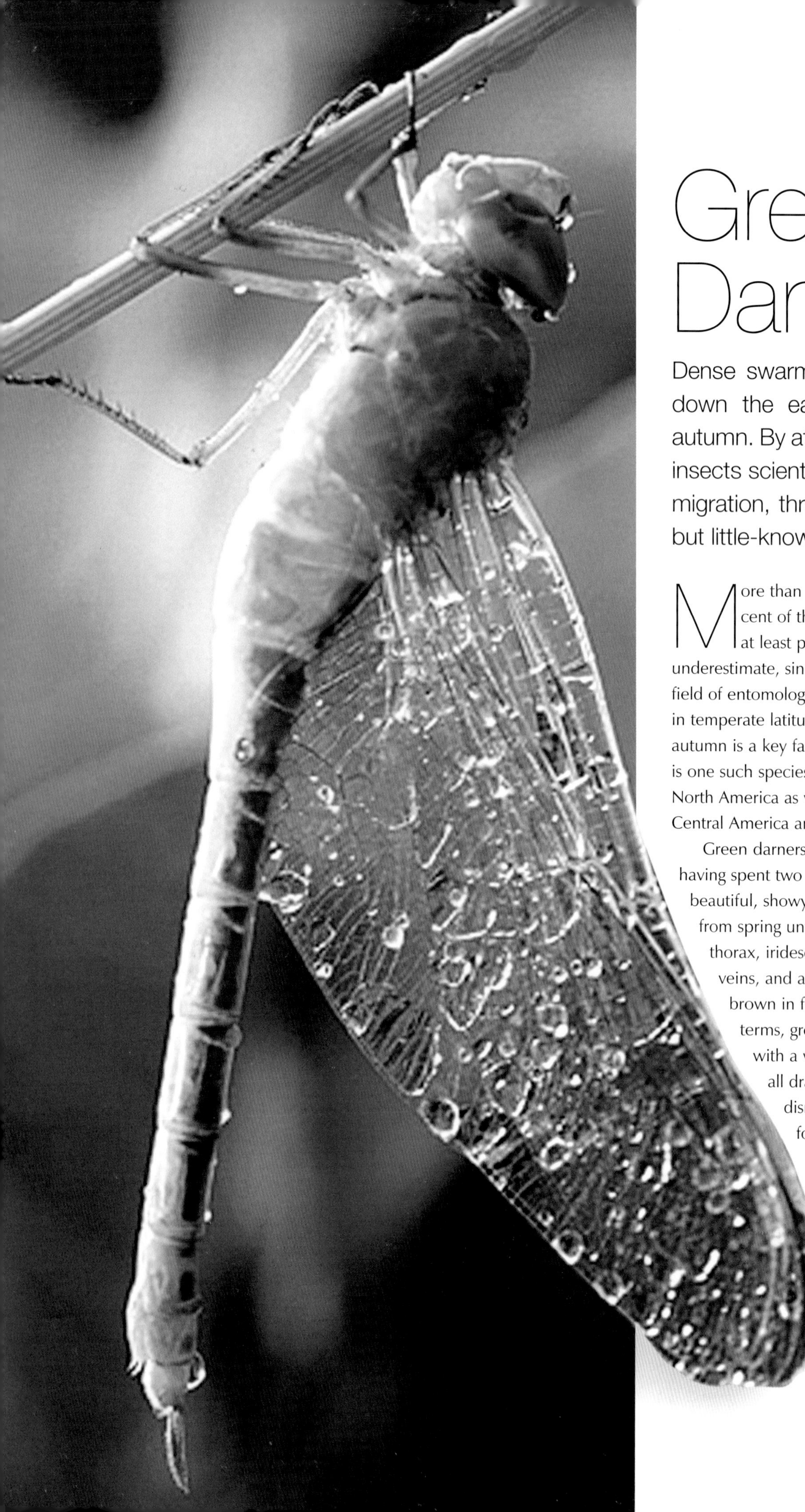

Green Darner

Dense swarms of green darner dragonflies funnel down the eastern seaboard of the USA in the autumn. By attaching tiny radio transmitters to these insects scientists have for the first time tracked their migration, throwing new light onto this spectacular but little-known phenomenon.

More than 50 species of dragonfly, approximately one per cent of the world total, are believed to be migrants or at least partially migratory. This figure could well be an underestimate, since dragonfly migration remains a poorly understood field of entomology. Most of the known dragonfly migrants are found in temperate latitudes, where the date of the first hard frost in the autumn is a key factor influencing their migration. The green darner is one such species, with a range that incorporates most of temperate North America as well as tropical areas further south, including Central America and Caribbean islands.

Green darners live for between four and seven weeks as adults, having spent two or three years as aquatic nymphs, and these beautiful, showy invertebrates can be seen on the wing anytime from spring until the autumn. They have a bright green head and thorax, iridescent wings traced with a net-like pattern of black veins, and a slender abdomen, bluish in males and yellow-brown in females. They are medium-sized in dragonfly terms, growing to a length of 6.5–7.5 cm (2.5–3 inches) with a wingspan of about 11 cm (4.5 inches), and – like all dragonflies – are fierce predators that catch and dismember other insects in mid-air, zipping back and forth to patrol their stretch of pond or lake shore.

AUTUMN EXODUS

It is generally assumed that green darners on the wing during the summer will live and die in the same small area, unless there is a drastic change to their habitat (a drought, for example)

Left Green darners (this is a female) spend their short summers hunting mosquitoes and midges while avoiding the attention of hawks. Their colouring provides camouflage.

MIGRATION PROFILE

Scientific name	*Anax junius*
Migration	Part of North American population migrates south in autumn
Journey length	Not known
Where to watch	Atlantic coast of USA
When to go	September–October

GREEN DARNER MIGRATION
Breeding range **→ Autumn migration route**

that forces them to move. But at least some, and perhaps most, of the adults that emerge later in the year become strongly migratory. These dragonflies form impressive aggregations as they head in a general southerly direction. The close-knit swarms follow "leading lines" in the landscape, such as ridges, cliffs, river valleys, the edges of lakes, and coasts, sometimes reaching near-plague proportions. One September day in 1992, an estimated 400,000 dragonflies flew over Cape May Point, New Jersey, USA in a mere 75 minutes. The majority of them were green darners.

In the USA, the best places to witness dragonfly migration in the fall include the Catskill and Appalachian mountains, the Great Lakes, and the Atlantic coast. The largest movements occur when a cold front sweeps through an area, probably because cold, northerly winds boost the dragonflies' progress south. The cue for their departure may be a sudden drop in night-time temperature – if it gets progressively colder for two nights, it is a good indication that favourable winds are on the way.

A pioneering radio-tracking study of green darners moving along the USA's East Coast has established that they behave much like birds on migration. The insects were found to pause regularly to rest and replenish their fat reserves, and avoided travelling on the windiest days when they were likely to be blown off course. According to the study's data, green darners could be capable of advancing over 700 km (450 miles) during their autumn migration, although this theory cannot be tested until long-range miniature radio transmitters are developed.

GENERATIONAL RELAY

The migration of green darners is one-way: the adults fly south never to return, and presumably die having laid their eggs. Green darners observed heading north in spring have relatively clean, unmarked wings, which means they must be a new generation of freshly metamorphosed adults, as opposed to survivors from the previous year's generation. If the dragonflies that migrated south somehow did manage to overwinter as adults and then return north, their wings would show signs of wear and tear.

Above The dragonflies are amazingly strong and manoeuvrable fliers for their size, due to their muscle-packed thorax and four wings that can beat in many different patterns.

RADIO-TRACKING DRAGONFLIES

Anyone who attempts to track the migration of individual insects in real time faces a logistical nightmare due to the small proportions of their study subjects. A green darner dragonfly weighs at most 1.5 g (0.05 oz) and a transmitter needs to be significantly lighter than this or the insect would be unable to migrate normally. In 2005, a team from Princeton University created an electronic tracking device weighing only 300 mg (0.01 oz), which can be attached to the underside of a dragonfly's thorax using a tiny drop of eyelash adhesive mixed with superglue. Each transmitter produces two radio pulses per second, and its signal is tracked on foot, in vehicles, or from a low-flying Cessna plane equipped with external antennas. A major drawback with this survey technique is the transmitters' short battery life, which so far has limited the period over which a dragonfly can be followed to no more than 10 days.

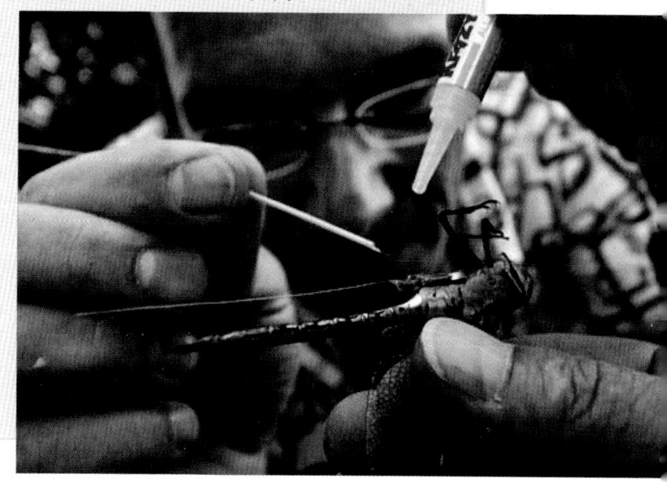

Right Great care has to be taken when attaching the miniature transmitters.

Desert Locust

Every so often the population of desert locusts explodes. The insects mutate into a voracious migrating form and launch a mass invasion of the surrounding area. Locust swarms rapidly reach plague proportions, reducing entire fields to a forest of bare stalks.

The biomass (weight of bodies) in a swarm of African desert locusts is staggering. Some of the largest swarms are estimated to contain 50,000 million individuals, cover an area of up to 1,000 square km (400 square miles), and take about six hours to pass overhead. In a single day locust plagues of this size can devour enough crops to feed 500 people for a year. Even more astonishing, the vast insect armies often seem to appear out of nowhere, in a matter of days. It is notoriously difficult to predict when invasions of locusts will happen, but they are far from random events. The outbreaks are caused by the locusts' opportunist life history, which leads to a "boom and bust" population structure characterized by sudden increases at irregular intervals.

SWARMING GRASSHOPPERS

Strictly speaking, the term "locust" describes about 10 species of grasshoppers in the family Acrididae that form periodic swarms, although various other kinds of grasshoppers also undergo population explosions and are sometimes referred to as locusts. True locusts are found in hot, arid parts of the world, especially in the tropics. A number of them have the capacity to wreak havoc – chief culprits include the Australian plague locust, the American locust, and the migratory and desert locusts, both of which occur in a huge region stretching from sub-Saharan Africa, east through the Middle East to southern Asia. However, what sets desert locusts apart is the stupendous scale of their swarms, their ability to travel great distances, and their potential for inflicting immense damage on agriculture and the livelihoods of millions of people.

Despite their name, desert locusts are not too fussy about their habitat. They thrive in a wide range of open or semi-open country, especially savanna and scrub. During droughts or years of average rainfall, they tend to be shy and unobtrusive insects, so are hardly noticed by local people. Rain changes everything.

SPLIT PERSONALITIES

Heavy downpours create a flush of new plant growth, prompting female locusts to lay eggs in the damp soil. Further rainfall triggers the eggs to hatch, and if there is an ideal combination of rain, temperature, and fresh vegetation, enormous numbers of eggs will be laid and hatch successfully. At first the nymphs (young locusts) are wingless eating machines, called hoppers. They consume their own weight in plant matter every 24 hours and develop quickly, moulting their tough exoskeleton five times to permit growth. At the fifth moult, they acquire wings and become adults. The complete life-cycle, from egg-laying to sexual maturity, takes as little as 45 days.

In their quiet periods, known as recessions, desert locusts are spread thinly, but sustained rains enable them to breed continuously. The resulting overcrowding drives them to mutate (see box). Within a few hours, the jostling packs of hoppers change their appearance and behaviour radically, morphing from plain green insects into sociable, boldly coloured ones with bright yellow, orange, and black stripes.

Below Thick clouds of locusts migrate with the wind, leaving behind a trail of devastation.

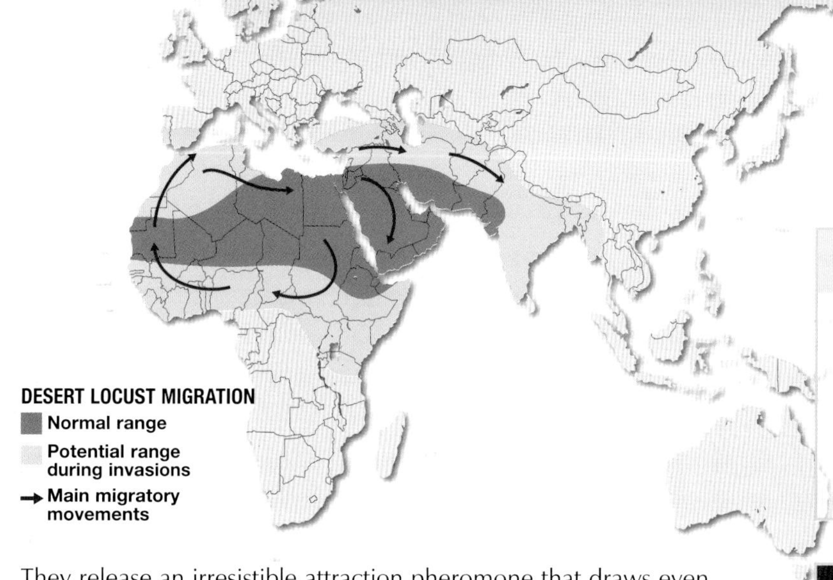

MIGRATION PROFILE

Scientific name	*Schistocerca gregaria*
Migration	Periodic eruption of migrating swarms
Journey length	Up to several thousand miles
Where to watch	West Africa
When to go	Locust invasions are irregular events

They release an irresistible attraction pheromone that draws even more hoppers from the surrounding area, so that their numbers start to grow exponentially.

When so-called "gregarious" hoppers mature, they look and behave differently, too. Normal adult locusts are brownish, nocturnal, and loners by nature, but these hoppers turn into yellow, day-active, crowd-loving adults with longer wings. Before long, they take off in a dense cloud that moves with a single purpose, in search of new areas to lay eggs. All of their offspring will automatically be of the swarming variety.

BLOWING IN THE WIND

Carried aloft by the wind, locust swarms are able to make amazing long-range journeys. Sometimes they are swept across the Mediterranean into southern Europe, over the Red Sea to the Arabian Peninsula, or hundreds of miles out into the Atlantic. In October 1988, a swarm of desert locusts crossed from West Africa to the Caribbean – a feat that required 4–6 days of flight time. In laboratory tests, locusts have flown up to 20 hours non-stop, which means any extra flight time is dependent on wind assistance.

The worst plagues of desert locusts can invade up to 30 million sq km (12 million square miles) of Africa and Asia, equivalent to about 20 per cent of the Earth's land surface, and cause destruction on a near biblical scale. In the end, the swarms always die out, due to exhausted food supplies, predation, disease, or adverse weather.

Below Egypt's pyramids at Giza, one of the great achievements of humankind, look insignificant compared to this huge swarm of locusts, which invaded North Africa in 2004.

PHASE CHANGE

How and why do young desert locusts transform from their solitary phase into their highly social migratory phase? Led by biologist Stephen Simpson, a team of researchers at Oxford University set out to establish if the phase change is triggered by visual clues, chemical signals such as scent, physical contact, or these and other factors working together. They found that, while sight and smell play a small part in the process, the dominant stimulus is touch. Population growth reduces the amount of space available to locust nymphs, which brush against one another more frequently, and when this pushing and shoving reaches a critical level the insects mutate. The dramatic change is activated by an area of sensitive hairs on their hind legs, which has been christened the "G-spot" for its role in stimulating gregarization. Insecticides might one day be developed that can desensitize the G-spot, thereby preventing locusts from swarming.

Right This locust has mutated into its swarm phase, a change triggered by physical contact with other locusts.

Global Migration Hotspots

Animal migration is, without doubt, one of the most enthralling spectacles that nature has to offer. From huge flocks of waders swirling in the sky to thundering herds of antelope racing across the savanna, and from mysterious gatherings of ocean-going fish to waves of turtles hauling themselves across a nesting beach, migration in its many forms has an unrivalled capacity to delight and inspire us. This gazetteer is intended to be a showcase of some of the world's top locations to see migratory animals. Each site has been selected because it provides the opportunity to experience an amazing migration event, often at close range. Due to migration's seasonal nature, most of the sites are best visited at a specific time of year – but even then, nothing can be guaranteed. Patience and perseverance are key virtues of the wildlife watcher.

PACIFIC OCEAN

Covering a third the Earth's surface, the Pacific supports more animal migrations than anywhere else. Its seamounts and remote islands are visited by pelagic sharks, turtles, and seabirds, while whales calve in its warm tropical waters.

NORTH AMERICA

The Midwest's huge herds of bison and pronghorn antelope are long gone, but North America still has thrilling spectacles, including trekking polar bears, swarming monarch butterflies and free-tailed bats, and impressive flocks of many birds.

CENTRAL AND SOUTH AMERICA

Central America is a migratory corridor used by vast numbers of birds travelling between temperate breeding grounds to the north and wintering areas in the tropics. Sheltered inshore waters are a major feeding area for pelagic fish, such as whale sharks; further offshore, the Galapagos Islands host the unusual breeding migration of land iguanas. Patagonia, at the southernmost tip of South America, is a wildlife-rich wilderness visited by breeding penguins, sealions, and right whales.

ATLANTIC

Numerous seabirds, whales – and a few fish, such as tuna and sharks – migrate between the Atlantic's cold waters at high latitudes and its warmer tropical waters. Islands along the Mid-Atlantic Ridge are a nesting ground for turtles and albatrosses.

EUROPE

There are no large-scale movements of mammals in Europe, and birds are by a long way the most obvious migrants. Millions of passerines and birds of prey fly to Europe from Africa to breed, while in winter, waves of waders, thrushes, and waterfowl arrive from the Arctic. Other spectacular migrations include those of European eels and Atlantic salmon, and the return journey of green and loggerhead turtles to nesting beaches in the Mediterranean.

ASIA

In the far north, tundra provides a summer home for migrant waders and waterfowl, which fly here from as far away as Australasia, while the nutrient-rich waters that wash the northeast of the continent attract seabirds and whales. The rolling steppe at middle latitudes is roamed by nomadic grazers such as Mongolian and goitered gazelles and saiga antelopes. Southern wetlands and coasts welcome a host of migratory waders in winter.

AFRICA

The savanna that swathes much of sub-Saharan Africa witnesses large-scale migrations of antelope, zebra, buffalo, and elephant. Good viewing opportunities occur at the end of the dry season, as animals become concentrated at dwindling water sources. Perhaps the greatest wildlife spectacle is the wildebeest migration of the Serengeti–Mara ecosystem. The grassy plains seem to run on forever (Serengeti is a Masai word meaning "land of endless space") and are a photographer's paradise.

ANTARCTICA

One of the world's most stunning feats of endurance is the march of emperor penguins back and forth across the frozen continent. This gruelling marathon starts during the Antarctic winter – the penguins are easier to see later in the year.

AUSTRALASIA

Isolated from the rest of the world for 45 million years, Australia has evolved a unique fauna. Many of the species that live in its arid centre have highly nomadic lifestyles. Various Australian and New Zealand coasts are important breeding grounds for a range of migratory species, including whales, turtles, shearwaters, and albatrosses. Christmas Island is the famous venue for the eye-catching migration from land to sea and back again of millions of red land crabs.

NORTH AMERICA

1. ARCTIC NATIONAL WILDLIFE REFUGE, ALASKA, USA
Main species Caribou, polar bear, snow goose, Dall sheep
When to go June–July
Habitat Coast, tundra, mountains
Visitor tip You are most likely to find wildlife on a guided tour

2. ADAMS RIVER, BRITISH COLUMBIA, CANADA
Main species Sockeye salmon
When to go August–October
Habitat Major river, rising in the Columbia Mountains
Visitor tip Large sockeye runs occur every four years

3. CHURCHILL, MANITOBA, CANADA
Main species Polar bear, beluga whale
When to go October–November (bears), July–August (belugas)
Habitat Coast, tundra; Hudson Bay freezes in winter
Visitor tip The most accessible location to see polar bears; more than 3,000 belugas visit the Churchill River estuary in summer

4. CHEYENNE BOTTOMS, KANSAS, USA
Main species Sandhill crane, shorebirds
When to go March–April
Habitat Wetland
Visitor tip An important staging area for migratory birds

5. DELAWARE BAY, DELAWARE AND NEW JERSEY, USA
Main species Red knot, green darner dragonfly, shorebirds, hawks, wood-warblers
When to go Late May (knot), September–October (other species)
Habitat Seashore, coastal lagoon, marsh
Visitor tip Spectacular flocks of knot pause here in spring

6. SEA OF CORTEZ, BAJA CALIFORNIA, MEXICO
Main species Gray, humpback, and sperm whales
When to go January–March
Habitat Sheltered coastal waters
Visitor tip Experience close encounters with gray whales

7. EL ROSARIO RESERVE, MICHOACÁN, MEXICO
Main species Overwintering monarch butterfly
When to go February–early March
Habitat Montane fir forest
Visitor tip Best visited on a sunny day in the early morning

8. VERACRUZ CITY, MEXICO
Main species Swainson's hawk, turkey vulture, other birds of prey
When to go August–October
Habitat Urban areas, farmland
Visitor tip More than five million raptors stream over this viewpoint

CENTRAL AND SOUTH AMERICA

9. BAY ISLANDS, HONDURAS
Main species Whale shark, Caribbean spiny lobster, green and hawksbill turtles
When to go February–April
Habitat Coral reefs and atolls
Visitor tip Part of the Western Hemisphere's longest barrier reef

10. LOS LLANOS, VENEZUELA
Main species Herons, ibises, storks, whistling-ducks
When to go November–March
Habitat Savanna, wetland
Visitor tip A haven for waterbirds, particularly in the dry season

11. PANTANAL, BRAZIL
Main species Yacare caiman, egrets, herons; wide variety of migratory freshwater fish
When to go August–October
Habitat Seasonally flooded swamp, woodland, and savanna
Visitor tip Fish move upriver to reproduce, sustaining enormous numbers of caiman and waterbirds

12. PENINSULA VALDES, ARGENTINA
Main species Magellanic penguin, southern sealion, southern elephant seal, southern right whale, orca
When to go November–January
Habitat Shingle beach
Visitor tip Amazing concentrations of breeding marine wildlife; Punta Tombo is another superb site

PACIFIC OCEAN

13. HAWAII, USA
Main species Humpback whale, green turtle
When to go January–March (humpbacks), all year (turtles)
Habitat Shallow coastal waters
Visitor tip Important nursery for humpbacks; green turtles sometimes come ashore to bask

14. GALAPAGOS ISLANDS, ECUADOR
Main species Land iguana
When to go June–July
Habitat Volcanic lava fields, scrub
Visitor tip Rare opportunity to observe migration in lizards

15. COCOS ISLAND, COSTA RICA
Main species Scalloped hammerhead shark, other large pelagic fish
When to go June–August
Habitat Seamount with powerful upwelling currents
Visitor tip Unrivalled opportunities to swim among schooling sharks

ATLANTIC OCEAN

16. ASCENSION ISLAND, UK
Main species Green turtle
When to go January–April (adults), March–June (hatchlings)
Habitat Sandy beach
Visitor tip Nesting females use almost every suitable area of island

17. SOUTH GEORGIA, UK
Main species Southern elephant seal, Antarctic fur seal, wandering albatross, king penguin
When to go November–February
Habitat Subantarctic islands
Visitor tip Teems with abundant, approachable wildlife

AFRICA

18. BANC D'ARGUIN NATIONAL PARK, MAURITANIA
Main species Shorebirds, terns
When to go October–February
Habitat Intertidal mudflats, sand-bars, small islands
Visitor tip Crucial wintering and staging area for migratory birds

19. SERENGETI–MARA, TANZANIA AND KENYA
Main species Wildebeest, plains zebra, Thomson's gazelle
When to go January–March (wildebeest calving), June–August (wildebeest cross Grumeti and Mara rivers)
Habitat Savanna
Visitor tip A multilodge safari gives the best chance of keeping up with migrating wildlife

20. KASANKA NATIONAL PARK, ZAMBIA
Main species Straw-colored fruit bat
When to go November–December
Habitat Mushitu swamp forest
Visitor tip Activity at the bat roosts is highest at dusk and dawn

21. OKAVANGO DELTA, BOTSWANA
Main species African elephant, buffalo, plains zebra, waterbirds
When to go May–June
Habitat Floodplain wetland
Visitor tip Mammals have young and birds are breeding at this time

22. KRUGER NATIONAL PARK, SOUTH AFRICA
Main species African elephant, impala, plains zebra, white rhino
When to go May–September
Habitat Savanna, thorn bush, woodland
Visitor tip The dry season offers best game viewing because vegetation thins

EUROPE

23. TARIFA COAST, SPAIN
Main species European white stork, birds of prey, passerines
When to go March–April, August–September
Habitat Beach, cliff, headland
Visitor tip Migrating passerines are most evident at first light as they frantically feed to refuel; large soaring birds move during the heat of the day

24. THE WASH, LINCOLNSHIRE AND NORFOLK, UK
Main species Red knot, dunlin, pink-footed goose
When to go November–March
Habitat Intertidal mudflats, salt-marsh
Visitor tip At high tide dense flocks of shorebirds crowd on the shore

25. WADDEN SEA, NETHERLANDS, GERMANY, AND DENMARK
Main species Ducks, geese, shorebirds
When to go All year
Habitat Shallow coastal waters, intertidal mudflats, salt-marsh
Visitor tip Several million waterfowl and shorebirds overwinter here

26. DANUBE DELTA, ROMANIA
Main species Dalmatian pelican, red-breasted goose
When to go April–June (breeding waterbirds), November–March (wintering waterfowl)
Habitat River, lake, reed-swamp
Visitor tip Europe's largest wetland throngs with birds in summer and winter

ASIA

27. HEMIS NATIONAL PARK, LADAKH, INDIA
Main species Snow leopard, wild sheep and goats
When to go January–February
Habitat Mountains
Visitor tip The park's contains about 100 elusive leopards

28. KEOLADEO NATIONAL PARK, RAJASTHAN, INDIA
Main species Ducks, geese, cranes, birds of prey
When to go October–March
Habitat Marsh, woodland, scrub
Visitor tip A magnet for wintering waterfowl from further north; many raptors visit, too

29. MAI PO, CHINA
Main species Shorebirds (up to 25 species)
When to go April–May
Habitat Mudflats, mangroves, shrimp ponds
Visitor tip Hosts migratory shorebirds en route to the Arctic

30. KAMCHATKA PENINSULA, RUSSIA
Main species Short-tailed shearwater, auks, gray whale
When to go June–July
Habitat Rocky coast, inshore waters
Visitor tip The productive seas off this volcanic peninsula support millions of seabirds and a variety of cetaceans in summer

31. SIPADAN ISLAND, BORNEO, MALAYSIA
Main species Green and hawksbill turtles
When to go July–August
Habitat Coral reef
Visitor tip Dozens of turtles can be seen on a single dive at this site

AUSTRALASIA AND ANTARCTICA

32. CHRISTMAS ISLAND, AUSTRALIA
Main species Red crab
When to go November–January
Habitat Rocky shore
Visitor tip Updates on crab movements are released via local radio

33. KAKADU NATIONAL PARK, NORTHERN TERRITORY, AUSTRALIA
Main species Magpie goose, Australian pelican, saltwater crocodile
When to go July–August
Habitat Floodplain wetland
Visitor tip The dry season offers the best wildlife watching

34. NINGALOO REEF, WESTERN AUSTRALIA, AUSTRALIA
Main species Whale shark, manta ray
When to go March–April
Habitat Coral reef
Visitor tip It is possible to snorkel with these giant filter-feeders

35. KAIKOURA, SOUTH ISLAND, NEW ZEALAND
Main species Albatrosses, sperm whale
When to go June–August
Habitat Deep ocean trench close to shore
Visitor tip A reliable base to see albatrosses and sperm whales

36. ANTARCTIC PENINSULA
Main species Emperor, Adélie, and gentoo penguins, various albatrosses and petrels, humpback whale
When to go December–January
Habitat Plankton-rich polar waters
Visitor tip Cruise ships often incorporate visits to South Georgia, the South Orkneys, and South Shetland Islands

Glossary

altitudinal migration The seasonal movement of an animal between a high- and low-altitude area; the movement is sometimes known as *vertical migration*.

anadromous A fish that lives mostly in the sea and travels to fresh water to reproduce. See also *catadromous*.

austral Of or relating to the south or the Southern Hemisphere; for example, the austral summer.

baleen A flexible, horny substance hanging in fringed plates from the upper jaw of baleen whales, which is used to strain plankton from seawater when feeding; also known as whalebone.

biological clock An internal "clock" that exists in almost all organisms, from those as tiny as bread mould to humans; its function is to time physiological and behavioural processes.

boreal Of or relating to forest areas of the temperate zone in the far north, dominated by coniferous trees such as fir, spruce, and pine.

broad front Describes migration in which widely dispersed migrants travel without deviating from their preferred directions, with no apparent detours caused by the landscape or other features. See also *narrow front*.

catadromous A fish that lives mostly in fresh water and travels to the sea to reproduce. See also *anadromous*.

cetacean A member of the mostly marine mammal order Cetacea, including the whales, dolphins, and porpoises.

circadian rhythm A regular cycle of bodily activities and functions that, in the absence of external clues, repeats every 24 hours or so. See also *biological clock*; *circannual rhythm*.

circannual rhythm An annual cycle of bodily activities and functions; it may or may not be influenced by external clues, such as day-length. See also *biological clock*; *circadian rhythm*.

dispersal In ecology, the movement of an animal from its current area, usually with no fixed direction or distance.

emigration Dispersal or migration of organisms away from an area.

eruption A large-scale *emigration* away from a particular region; also known as an "irruption" or "invasion".

escape movement A mass movement from an area that has suddenly become extremely inhospitable, due, for example, to a storm or heavy snowfall.

flyway An invisible migratory corridor, or air route, used year after year by migratory birds, insects, or bats.

geo-magnetic sense The ability of some animals to detect magnetic fields, used in orientation.

great circle route The shortest path between two points on the Earth's surface; it involves a constant, progressive change of direction during the journey.

hyperphagia Intensive feeding prior to migration, in order to lay down fat reserves as fuel.

jet stream A fast-flowing air current found at high altitude.

latitudinal migration Migration between higher and lower latitudes.

leading line A topographical feature, such as coastline, lake shore, or mountain ridge, along which migrants tend to travel.

magnetic compass A compass based on the Earth's magnetic field.

migrant Any organism that takes part in a migration.

migration Travel with a clear purpose from one area to another, often following a well-defined route to a familiar destination, and often at a specific season or time; migration may occur once or many times in the lifetime of an organism.

migratory divide An imaginary line on a map separating two or more breeding populations that migrate in different directions; on one side of the line migrants travel in one direction (for example, southwest), and on the other side they move in another direction (for example, southeast).

moult migration Movement to a special area to moult, seen in certain birds, reptiles, and marine mammals.

narrow front Describes migration in which migrants from a wide area are concentrated to pass along coasts and peninsulas, or through narrow valleys or other geographical features. See also *broad front*.

navigation Movement along a specific course in order to reach a particular goal; it requires knowledge of the distance between the present location and the destination. See also *orientation*.

nomadic Describes an animal that travels widely but with no fixed directional preferences or migratory schedule.

orientation Movement in a certain direction, using a variety of external clues to keep on the desired compass heading. See also *navigation*.

partial migrant A species in which not all individuals regularly migrate; only those from a particular area, or of a certain age or sex, are migratory.

passerine A member of the order Passeriformes, which includes more than half of all bird species; sometimes known collectively as the perching birds, or, less accurately, as songbirds.

PAT Pop-up archival tag. A data logger attached to an aquatic animal that stores information to be retrieved later; generally used to study the movements of marine fish and turtles.

pelagic Ocean-dwelling; usually describes wide-ranging marine organisms that occur far from land.

philopatry The tendency of an animal to stay in, or return to, a certain location in order to breed or feed.

phonotaxis Orientation towards a sound.

plankton A collective term for the small or microscopic organisms, including plants, algae, bacteria, protozoans, crustaceans, and the larvae of larger animals, that float or drift in immense numbers in both fresh and salt water, especially near the surface.

PTT Platform transmitter terminal. A satellite-tracking device that is attached to an animal to determine its position; the appliance emits a signal that is picked up by the ARGOS array of satellites orbiting the Earth.

raptor A diurnal bird of prey, such as a falcon, hawk, eagle, or osprey.

radio-tracking The location and tracking of a migratory animal that has been marked with a radio transmitter.

resident Describes an individual or population that remains all year round in the same area; also known as non-migratory or sedentary.

ringing The research technique of placing a metal ring carrying a serial code and return address on the leg of a bird; if the bird is caught or found dead, its migratory movements can be tracked.

satellite telemetry A term, derived from the ancient Greek for "distance measuring", which refers to the research technique of using a *PTT* to track an animal's movements.

sedentary *see* resident

staging area A place where large numbers of migrants pause during their migratory journey, usually to rest and refuel but sometimes also to moult; also known as a staging post or stopover site.

star compass A compass based on the position of stellar features in the night sky, which depends on an animal being able to detect the rotation of patterns of stars around a fixed point.

Sun compass A compass based on the position of the Sun in the sky.

tubenose A bird in the marine order Procellariiformes, including the albatrosses, shearwaters, and petrels; the term derives from the prominent tubular nasal passages on the bill.

vagrant An individual that has reached an area outside its normal range and migration route, due to a navigational error or unfavourable conditions encountered en route.

vertical migration The regular, daily, or seasonal, movement of an animal between a higher and lower area, in mountains or a body of water such as a sea or lake; in mountains, also known as *altitudinal migration*.

zugunruhe A German word meaning "migratory restlessness"; describes the agitated state of migratory birds immediately prior to their journey; best known in *passerines*.

Index

Numbers in **bold** indicate pages where an in-depth treatment is given; numbers in *italics* indicate images.

T

U

V

W

Y

Z

Acknowledgements

A lot of people have contributed their time and expertise to this book. For their insightful comments on the text, I would like to thank Jonathan Elphick, Ian Redmond, Rob Houston, David Burnie, and Emily Bennitt (Mammal Research Unit, University of Bristol). George Schaller (Wildlife Conservation Society) provided valuable information about Mongolian gazelles, Andy Elliott (editor, *Handbook of the Birds of the World*, Lynx Edicions) kindly answered my questions about carmine bee-eater taxonomy, and Dave Roberts (Manitoba Conservation) shared his knowledge of migration in red-sided garter snakes. Scott Shaffer (Department of Ecology and Evolutionary Biology, University of California Santa Cruz) drew my attention to the various ways in which tracking and data logging technology is being applied to the study of seabird migration, and provided the diagrams reproduced on page 37. I would also like to thank the experts of University California Press and the Natural History Museum, London, who have saved me from innumerable errors. The staff of the Zoological Society of London's wonderful library have been, as ever, exceptionally helpful, and everyone at *BBC Wildlife* magazine has given much support. This book would not have been possible without the skill and dedication of the Marshall Editions team, including Amy Head, Deborah Hercun, Ivo Marloh and—above all—Paul Docherty. Finally, I would like to thank my wife, Louise, for her love and not inconsiderable patience.

Picture Credits

Marshall Editions would like to thank the following for their kind permission to reproduce their images:

KEY

FLPA/HS =	Frank Lane Picture Agency/Holt Studios
FLPA/IB =	Frank Lane Picture Agency/Imagebroker
FLPA/MP =	Frank Lane Picture Agency/Minden Pictures
FLPA =	Frank Lane Picture Agency
NOAA =	National Oceanic Atmospheric Administration
NPL =	Nature Picture Library
P/OSF =	Photolibrary/Oxford Scientific Films
PH/NHPA =	Photoshot/NHPA
USFW =	US Fish and Wildlife Service

t = top, **b** = bottom, **c** = centre, **r** = right, **l** = left

Pages: 2–3 Corbis/Tony Wilson-Bligh; **4–5** Corbis/Stuart Westmorland; **6** Corbis/Jonathan Blair; **8–9** FLPA/IB; **10** NHPA/John Shaw; **11b** Corbis/Paul Souders; **11tr** Corbis/Ron Sanford; **11br** Corbis/Bettman Archives; **12** NASA; **13t** Corbis/Nick Garbutt; **13b** FLPA/Flip Nicklin/MP; **14** NPL/Michael D Kern; **15t** Alamy/Visual & Written SL; **15b** NPL/Dietmar Nill; **16** P/OSF/Martyn Colbeck; **17tl** Corbis/Romeo Ranoco; **171tr** FLPA/Francois Merlet; **18** NPL/Markus Varesvuo; **19t** Corbis/Paul A Souders; **19b** NPL/John Waters; **20–21** Corbis/DLILLC; **21tl** Corbis/Hinrich Baesemann/dpa; **22** NPL/Doug Perrine; **23** Corbis; **24** Ardea/George Reszeter; **25** P/OSF/Doug Allan; **27** NPL/Dietmar Nill; **28** FLPA/MP/Michael Durham; **29** PH/NHPA/Michael Patrick O'Neill; **30–31b** FLPA/Martin B Withers; **31t** FLPA/Reinhard Dirscherl; **32** P/OSF/Thorsten Milse; **33tl** Mary Evans Picture Library; **33b** Alamy/Imagebroker; **34** NPL/Andy Sands; **35** NPL/Doug Perrine; **36** NOAA/G. De Metrio; **37tl** FLPA/MP/Cyril Ruoso; **37bl** USFW/Togiak National Wildlife Refuge/Gail Collins; **37r** Scott Shaffer et al (PNAS, August 22 2006, vol. 103 no. 34) © 2006 National Academy of Sciences, USA; **38** P/Craig Aurness; **39t** Corbis/Transtock; **39cr** PH/NHPA/George Bernard; **39b** Corbis/Du Huaju/Xinhua Press; **40–41**Corbs/Jeffrey Arguedas/epa; **41tr** NPL Andrey Zvoznikov; **42–43** Corbis/Winfried Wisniewski; **44–45** FLPA/MP/Michio Hoshino; **45t** Corbis/Theo Allofs; **45b** Corbis/Zefa/Alan & Sandy Carey; **46** Corbis/Hans Strand; **47t** Alamy/Bryan & Cherry Alexander Photography; **47b** Corbis/Daniel J Cox; **48–49** FLPA/MP/Colin Monteath; **49cl** FLPA/MP/Zhinong Xil; **49cr** Corbis/Daniel J Cox; **50** Corbis/Paul A Souders; **51b** FLPA/MP/Michael Mauro; **52** FLPA/MP/Jim Brandenburg; **53b** Corbis/Bettmann; **54–55** Ardea/Robyn Stewart; **54cl** Corbis/Karl Ammann; **54cr** Corbis/Nigel J Dennis; **55** Corbis/Martin Harvey; **56–57** Corbis/Peter Johnson; **57br** NPL/Anup Shah; **58** Corbis/Tim Davis; **59** NPL/Tony Heald; **60–61** Corbis/Frans Lanting; **61bl** Corbis/Yann Arthus-Bertrand; **60–61bc** FLPA/ Frans Lanting; **61t** Corbis; **61br** Corbis/Kevin Schafer; **62** NPL/Gertrud & Helmut Denzau; **63** Alamy/John Warburton-Lee Photography; **63b** George Schaller/Wildlife Conservation Society, New York; **64** NPL/Solvin Zankl; **65** Ardea/M Watson; **65b** NPL/Solvin Zankl; **66–67** Corbis/Paul Souders; **67t** Corbis/Frans Lanting; **68** Ardea/Francois Gohier; **69t** Corbis/Michael S Yamashita; **69b** Alamy/Alexandra Morrison; **70** Ardea/D Parer & E Parer; **71t** FLPA/Tui De Roy; **71b** FLPA/MP/Pete Oxford; **72** Corbis/Zefa/Markus Botzek; **74** Corbis/Roger Garwood & Trish Ainslie; **75** NPL/Jurgen Freund; **76–77** Corbis/Stuart Westmorland; **78** Corbis/Paul A Souders; **78b** FLPA/MP/Flip Nicklin; **79** Rex Features/Phil Rees; **80** FLPA/Flip Nicklin; **81** P/OSF/Gerard Soury; **82–83** Alamy/Mark Conlin; **83br** FLPA/Flip Nicklin; **84** Corbis/Staffan Widstrand; **85t** Corbis/Dan Guravich; **85b** Corbis/Dan Guravich; **86–87** Corbis/Theo Allofs; **87cr** FLPA/MP/Flip De Nooyer; **88** PH/NHPA/Michael Patrick O'Neill; **89t** NPL/Doug Perrine; **89b** NPL/Doug Perrine; **90** FLPA/IB/J W Alker; **91bc** PH/NHPA/Kevin Schafer; **91br** NPL/Jurgen Freund; **92–93** Sea Pics, Hawaii/Paul Humann; **93bl** NPL/Doug Perrine; **93bc** NPL/Doug Perrine; **93br** Dominguez S. Malavieille J. & Lallemand S. E.: Deformation of accretionary wedges in response to seamount subduction – insights from sandbox experiments; Tectonics, 19, 1, 182–196, 2000; **94–95** Corbis/Louie Psihoyos; **94b** NPL/Jurgen Freund**;** **95t** Getty Images/National Geographic/Brian J Skerry; **96** Corbis/Jeffrey L Rotman; **97** Corbis/Zefa/Tobias Bernhard; **98** Alamy/Visual & Written SL; **99** Alamy/CuboImages srl; **100** FLPA/MP/Jim Brandenburg; **100tc** NPL/Michel Roggo; **101** Corbis/Sanford/Agliolo; **102** NHPA/Daniel Heuclin; **102t** P/Rodger Jackman; **102b** FLPA/Terry Whittaker; **103–104** Ardea/Ken Lucas; **104b** NPL/Jeff Rotman; **105bl** FLPA/MP/Norbert Wu; **105br** FLPA/D. P. Wilson; **106** Getty Images/David Tipling; **107br** FLPA/MP/Flip Nicklin; **108–109** P/OSF/Perrine Doug; **109t** PH/NHPA/Linda Pitkin; **110** NPL/Georgette Douwma; **111t** FLPA/MP/Fred Bavendam; **111b** NPL/Chris Gomersall;**112–113** Corbis/Layne Kennedy; **114** NPL/Rolf Nussbaumer; **115** FLPA/Fritz Polking; **116–117** Alamy/Malcolm Schuyl; **117tr** FLPA/HS/Chris & Tilde Stuart; **118** NPL/Tom Vezo; **119** PH/NHPA/Brian Hawes; **120** Corbis/Ralph A Clevenger; **121t** Corbis/Darrell Gulin; **121b** Ardea/Ian Beames; **122–123** FLPA/MP/Hiroya Minakuchi; **123cr** Alamy/Steve Bloom Images; **123br** NPL/Chris Gomersall; **124** PH/NHPA/Dave Watts; **125** NPL/Nature Production; **126** Alamy/Chris Gomersall; **127t** NPL/Chris Gomersall; **127b** FLPA/Robert Canis; **128–129** Ardea/Duncan Usher; **129l** P/OSF/Konrad Wothe;**130t** FLPA/Dickie Duckett; **130b** Corbis/Lech MuszyÒski; **131** FLPA/MP/Jim Brandenburg; **132** PH/NHPA/Kevin Schafer; **132cl** Getty Images/Stone/Will & Deni McIntyre; **133** FLPA/MP/Tom Vezo; **134–135** Corbis/Theo Allofs; **135bc** FLPA/John Watkins; **136** NPL /Tom Hugh-Jones; **137** FLPA/MP/Yva Momatiuk/John Eastcott; **138t** PH/NHPA/Stephen Krasemann; **138b** FLPA/Mark Sisson; **139** Corbis/Arthur Morris; **140** PH/NHPA/Roger Tidman; **141** PH/NHPA/Jari Peltomaki; **142** FLPA/Mark Sisson; **143t** FLPA/Malcolm Schuyl; **143b** FLPA/MP/Michael Quinton; **144–145** NPL/Vincent Munier; **145c** FLPA/Malcolm Schuyl; **145b** PH/NHPA/Andy Rouse; **145t** FLPA/S & D & K Maslowski; **146** P/OSF; **147t** Alamy/Tom Uhlman; **147b** P/OSF/Daybreak Imagery;**148–149** FLPA/Philip Perry; **148b** FLPA/MP/Mitsuaki Iwago; **150** NPL/Kim Taylor; **151** NPL/John Downer; **152** PH/NHPA/Alan Barnes; **153t** FLPA/HS/Mike Lane; **153b** Corbis/Remi Benali; **154** FLPA/Derek Middleton; **155l** FLPA/Derek Middleton; **155r** NPL/David Kjaer; **156t** FLPA/David Hosking; **156b** NOAA (http://apod.nasa.gov/apod/ap040903.html); **158–159** Corbis/Frans Lanting; **159t** FLPA/IB/André Skonieczny; **159bl** Rex Features/Finlayson/Newspix; **159bc** NPL/Larry Michael; **159br** FLPA/IB/Andreas Pollok; **161tr** PH/NHPA/T Kitchin & V Hurst; **161**Ardea/Jean Paul Ferrero;**162** P/OSF/Scott Camazine; **163** PH/NHPA/John Shaw; **163b** © Christian Ziegler/Smithsonian Tropical Research Institute, Panama; **164–165** Corbis/Pierre Holtz; **165cr** PH/NHPA/Stephen Dalton.